北京理工大学“双一流”建设精品出版工程

Fundamentals of Statistical Signal Processing

统计信号处理基础

贾丽娟 ◎ 编

北京理工大学出版社
BEIJING INSTITUTE OF TECHNOLOGY PRESS

图书在版编目（CIP）数据

统计信号处理基础 = Fundamentals of Statistical Signal Processing：英文 / 贾丽娟编. -- 北京：北京理工大学出版社，2023.2

ISBN 978-7-5763-2144-9

Ⅰ. ①统…　Ⅱ. ①贾…　Ⅲ. ①统计信号-信号处理-英文　Ⅳ. ①TN911.72

中国国家版本馆 CIP 数据核字（2023）第 034918 号

出版发行 / 北京理工大学出版社有限责任公司
社　　址 / 北京市海淀区中关村南大街 5 号
邮　　编 / 100081
电　　话 / (010) 68914775（总编室）
　　　　　(010) 82562903（教材售后服务热线）
　　　　　(010) 68944723（其他图书服务热线）
网　　址 / http://www.bitpress.com.cn
经　　销 / 全国各地新华书店
印　　刷 / 廊坊市印艺阁数字科技有限公司
开　　本 / 787 毫米×1092 毫米　1/16
印　　张 / 8.5
字　　数 / 200 千字
版　　次 / 2023 年 2 月第 1 版　2023 年 2 月第 1 次印刷
定　　价 / 49.00 元

责任编辑 / 王梦春
文案编辑 / 把明宇
责任校对 / 刘亚男
责任印制 / 李志强

Preface

This book, titled *Fundamentals of Statistical Signal Processing*, provides a systematic introduction to various theories and methods of stochastic signal processing. The content mainly includes discrete-time stochastic processes, signal modeling, spectrum estimation, Wiener filtering, Kalman filtering, and adaptive filtering. It aims to model, analyze, and process stochastic problems in engineering practice in a statistical viewpoint. Theories and methods discussed in this book have been widely used in engineering fields such as radar, communication, speech and image processing, automatic control, biomedicine, and so on.

This book is mainly used for the *Fundamentals of Statistical Signal Processing* (*Full English*) course for master students who have studied *Digital Signal Processing* and *Random Signal Analysis* courses. This book is also helpful as teaching and reference material for postgraduates who take courses related to statistical signal processing.

The author believes that in the development of China's technological powerhouse, we must lay a solid theoretical foundation and closely combine the research and development of the application foundation layer with the application layer to strengthen the drive for innovation and development. In recent years, the Party Central Committee and the State Council have attached great importance to information technology work. Informatization has entered a new stage of accelerating digital development and building a digital China. It should be noted that signal and information processing is a key link in the construction of social informatization, and is also a rapidly developing basic theoretical discipline. As the saying goes, "We should broaden our vision in the observation of the world." These are waiting for the hard work and intelligent creation of many aspiring individuals. At the same time, I hope that this book can play a role in revitalizing the country through science and education and achieving high-quality social development in the journey of realizing the great rejuvenation of the Chinese nation.

Contents

Chapter 1

Introduction

This book introduces the theory and algorithms used for the analysis and processing of random signals and their applications in practice. The accurate values of stochastic signals are unpredictable, but their average statistical properties exhibit regularity. In this way, we can describe random signals using statistical averages rather than explicit equations. Dealing with random signals, we mainly focus on the statistical description and modeling of the dependence between the values of signals, and their application to theoretical and practical problems.

Stochastic signals are described mathematically by probability theory, random variables, and stochastic processes. In practice, we deal with random signals by statistical techniques. In this book, we will introduce optimum signal processing methods which help to develop practical statistical signal processing techniques and evaluate their performance. In the last chapter, we introduce the knowledge of adaptive signal processing, which involves the use of optimum and statistical signal processing methods to design signal processing systems running in real time.

The purpose of this book is to provide a unified introduction to the theory, implementation, and applications of statistical signal processing. We focus on the key topics of signal modeling, spectral estimation, optimal filtering, and adaptive filtering, which we consider fundamental and have important applications. This book provides students with basic concepts and methodologies that lay a solid foundation for further study.

1.1 Discrete-Time Stochastic Processes

An introductory course in digital signal processing is concerned with the analysis and design of systems for processing deterministic discrete-time signals. A deterministic signal may be defined as one that can be described by a mathematical expression or that can be reproduced repeatedly. Examples of deterministic signals include the unit sample, a complex exponential, and the response of a digital filter to a given input.

However, it is necessary to consider a more general type of signal known as random process in engineering practice. A random process is an ensemble of signals that is defined by its statistical properties. For example, a random process may be a collection of all possible sinusoids with an invariable amplitude and frequency, and the randomness of this process is contained in its phase. Another more frequently considered random process is noise, which is ubiquitous. Noise may be quantization errors that occur in a quantizer or sampler, it may be round-off noise of a digital filter, or it may be the background noise picked up by a microphone or a radar antenna.

A discrete-time random process is an indexed sequence of random variables. The most important notions for describing these random variables include autocorrelation and power spectrum. The autocorrelation is a second-order statistical characterization of a random process, which indicates the independence between values of the sequence. It can be exploited to design systems for predicting a signal, which are important in applications such as speech compression, noise filtering, and signal detection. The power spectrum is the Fourier transform of the autocorrelation, which is the frequency domain description of random processes.

1.2 Signal Modeling

In many theoretical and practical applications, we are interested in generating random signals with certain properties or obtaining an efficient representation of certain random signals that possesses their correlation and spectral features. Signal model is a mathematical description which provides an efficient approximation of a signal.

In practical applications, we are interested in linear parametric models. A good linear parametric model includes these features: (1) the number of parameters should be as few as possible, (2) estimation of the parameters from the data is easy, and (3) the parameters are physically meaningful.

If we develop a successful parametric model for a signal, then we can use the model for multiple applications:

1. To better understand the physical mechanism generating the signal.
2. To track changes in the source of the signal and identify their cause.
3. To synthesize artificial signals like the natural ones.
4. To extract parameters for pattern recognition applications.
5. To get an efficient representation of signals for data compression.
6. To forecast future signal behavior.

In practice, signal modeling involves the following steps: (1) selecting an appropriate model, (2) select the right number of parameters, (3) fitting of the model to the actual data, and (4) testing the model in practical application.

1.3 Spectrum Estimation

The frequency domain provides a different viewpoint from which we can analyze a random process. The power spectrum is the Fourier transform of the autocorrelation of a stationary process. In lots of applications the power spectrum of a process is necessary to be known. For example, the IIR Wiener filter is defined by the power spectral densities of the input and output of that filter. The power spectrum is also important for detecting narrowband processes buried in noise.

If we simply Fourier transform signals of a stationary process, the result will not be a statistically reliable estimate of the power spectrum of that process. But if we can get the signal model of that process, we can use this model to estimate the spectrum. In this way, we will find that many methods developed for signal modeling are useful in solving the spectrum estimation problem.

1.4 Optimal Filters

Optimum filter is a filter that performs better than any other filter on certain criteria. For example, in the DSP curriculum for undergraduates, we have learned how to design a linear phase FIR filter that is optimum in the Chebyshev sense of minimizing the maximum error between the frequency responses of the designed filter and the ideal filter.

The Wiener and Kalman filters are also optimum filters. A Wiener filter is designed to process a given input, and form the best estimate of a desired response in the mean square sense. The Kalman filter may be used similarly to find the best estimate of a nonstationary process recursively. Both are used to solve a variety of problems including prediction, interpolation, deconvolution, and smoothing.

1.5 Adaptive Filters

The final topic considered in this book is adaptive filtering. Throughout chapters discussing signal modeling, spectrum estimation, and Wiener filtering, it is assumed that the processes are stationary, which means their statistical properties are time invariant. However, this would not be the case in the real world. In the last chapter of this book, these problems are reconsidered within the context of nonstationary processes.

The gradient descent algorithm can be used to solve the Wiener-Hopf equations and design the Wiener filter, which will be well behaved in terms of its convergence properties. But it is not generally used in practice, for it requires *a priori* knowledge of the statistical properties of the process.

The Least-Mean-Squares (LMS) algorithm is a stochastic gradient method which has been successfully used in many applications. Replacing the true gradient in a gradient descent algorithm with a gradient estimate, LMS is efficient in terms of its convergence properties. While the LMS algorithm is designed to solve the Wiener filtering problem by minimizing a mean square error, a deterministic least squares approach leads to the development of the RLS algorithm. Though RLS is much more computationally complex, it performs better than LMS algorithm. Both LMS and RLS are extensively used in applications such as linear prediction, channel equalization, interference cancelation, and system identification.

1.6 Organization of This Book

In this section, we provide an overview of the topics covered in this book to help the readers understand the interdependence among chapters (see Figure 1.1).

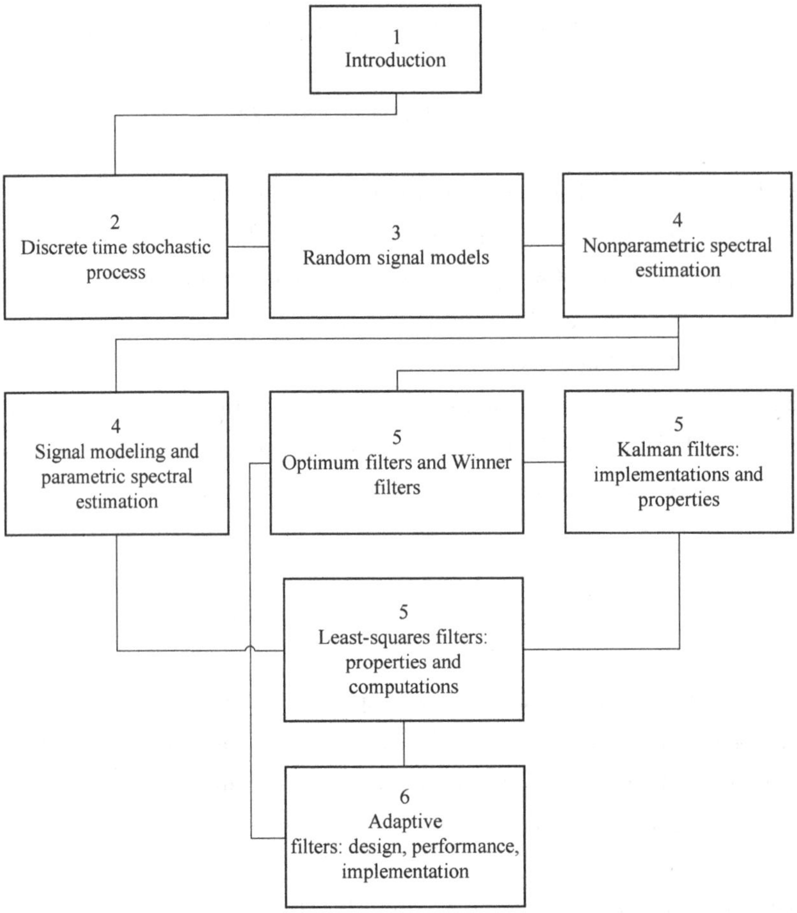

Figure 1.1 Flowchart organization of the book's chapters and their contents

In Chapter 2, we review the fundamental topics in the theory of random variables and random sequences and elaborate on certain crucial notions that are repeatedly used throughout the rest of the book. Chapter 3 provides a detailed study of the theoretical properties of signal models, assuming that the relevant signals can be modeled by random processes with known statistical properties. Chapter 4 presents some practical methods for nonparametric and parametric estimation of correlation and spectral densities. In Chapter 5, we focus on the design of optimum filters, including Wiener filter, least squares filter, and Kalman filter. Finally, Chapter 6 illustrates the adaptive filtering methods, presents the basic knowledge of LMS and RLS algorithms.

Chapter 2

Discrete-Time Stochastic Processes

2.1 Introduction

Signals may be classified into two types: deterministic signals and random signals. A deterministic signal can be reproduced exactly with repeated measurements, as exemplified by the unit impulse response of a linear shift-invariant system. On the other hand, a random signal is one that is not repeatable in a predictable manner, such as quantization noise produced by an analogue-to-digitalconversion(A/D) converter, background clutter in radar images, speckle noise in synthetic aperture radar (SAR) images, and engine noise in speech transmissions and so on. Although random signals are evolving in time in an unpredictable manner, their average properties can be often assumed to be deterministic; that is, they can be described by explicit mathematical formulas. This is the key for the modeling of a random signal as a stochastic process.

In this book, many of the signals we characterize and analyze are stochastic processes. Discrete-time stochastic processes are mainly described as an indexed sequence of random variables in this chapter. For a stochastic process, we define the characterization of its ensemble averages such as the mean, autocorrelation, and autocovariance, and explore the properties of these averages. Next the notion of stationarity and the concept of ergodicity are introduced. Then, we proceed to define the power spectrum of a stochastic process, which is the Fourier transform of the autocorrelation function. As an important characterization of a random process, the power spectrum will be used frequently in this book. Furthermore, we explore the effect of filtering a random process with a linear shift-invariant filter. Finally, an outline of an important class of random processes is described, including autoregressive (AR) and autoregressive moving average (ARMA) random processes, which may be generated by filtering white noise with a linear shift-invariant filter. In addition to the filtered white noise processes, we also look at harmonic processes that consist of a sum of random phase sinusoids or complex exponentials.

2.2 Definitions of Random Processes

Many natural sequences can be characterized as random signals because we cannot determine their values precisely, that is, they are unpredictable. A natural mathematical framework for the description of these discrete-time random signals is provided by discrete-time stochastic processes.

In this section, we consider the characterization and analysis of discrete−time random processes.

A discrete-time random process may be viewed as a mapping from the sample space Ω of experimental outcomes into the set of discrete-time signals $x(n)$. For example, as shown in Figure 2.1, for each experimental results ω_i , in the sample space Ω, there is a corresponding discrete-time signal $x_i(n)$. However, when describing and analyzing random processes, another perspective is often more practical. Specifically, note that for a particular value of n, that is $n = n_0$, the signal value $x(n_0)$ is a random variable defined in the space of the sample space Ω. In other words, for each $\omega \in \Omega$ there is a corresponding value of $x(n_0)$. Thus, a random process can also be thought of as an indexed sequence of random variables

$$\cdots, x(-2), x(-1), x(0), x(1), x(2), \cdots$$

where each random variable in the sequence has a probability distribution function

$$F_{x(n)}(\alpha) = \Pr\{x(n) \leqslant \alpha\}$$

and probability density function

$$f_{x(n)}(\alpha) = \frac{\mathrm{d}}{\mathrm{d}\alpha} F_{x(n)}(\alpha) \tag{2.1}$$

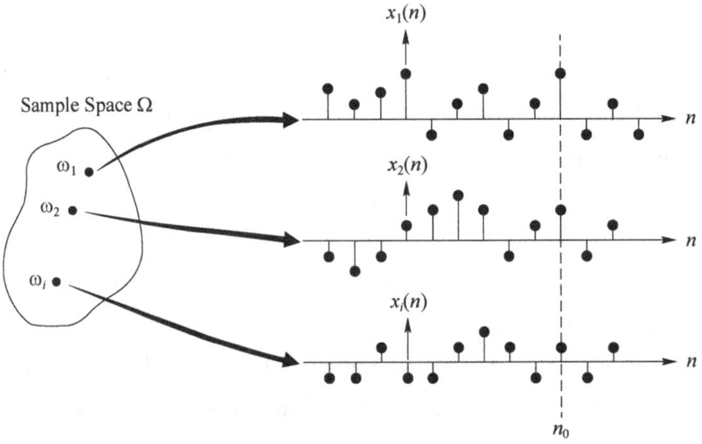

Figure 2.1 The generation of a random process from a sample space point of view

However, in order to form a complete statistical characterization of a random process, in addition to the first-order density function, the joint probability density or distribution functions that define how collections of random variables are related to each other must be defined. In particular, joint distribution functions are required

$$F_{x(n_1),\cdots,x(n_k)}(\alpha_1, \cdots, \alpha_k) = \Pr\{x(n_1) \leqslant \alpha_1, \cdots, x(n_k) \leqslant \alpha_k\}$$

for any collection of random variables $x(n_i)$. Significantly different types of random process can be generated depending on the specific shape of these joint distribution functions. For instance, a random process is formed from a sequence of Gaussian random variables $x(n)$. If the random variables $x(n)$ are uncorrelated, then the process, known as white Gaussian noise, is a very random, noise-like sequence. On the other hand, if $x(n) = \alpha$, where α is a Gaussian random variable, then each random process in the ensemble is equal to a constant for all n. Thus, though both the processes have the same first-order statistics, they are significantly different due to the differences between their higher-order density functions.

2.3 Time-Domain Statistical Characteristics

2.3.1 Ensemble Averages

Since a discrete-time random process is an indexed sequence of random variables, we can calculate the mean of each of these random variables and generate the deterministic sequence

$$m_x = E\{x(n)\} \tag{2.2}$$

which is called the mean of the process. Similarly, computing the variance for each random variable within the sequence

$$\sigma_x^2(n) = E\{|x(n) - m_x(n)|^2\} \tag{2.3}$$

defines the variance of the process. These first-order statistics represent ensemble averages and in general, both depend on n. The variance represents the mean average squared deviation of the process away from the mean while the mean defines the average value of the process as a function of n.

Two additional important ensemble averages in the study of random processes are the autocovariance

$$c_x(k,l) = E\{[x(k) - m_x(k)][x(l) - m_x(l)]^*\} \tag{2.4}$$

and the autocorrealation

$$r_x(k,l) = E\{x(k)x^*(l)\} \tag{2.5}$$

relating the random variables $x(k)$ and $x(l)$. Note that if $k = l$ then the auto covariance function reduces to the variance,

$$c_x(k,k) = \sigma_x^2(k)$$

If the product in equation (2.4) is extended, then it follows that the autocovariance and autocorrelation sequences and is related by

$$c_x(k,l) = r_x(k,l) - m_x(k)m_x^*(l)$$

Thus, for zero mean random processes, the auto-covariance and autocorrelation are equal. In this book, for convenience, unless stated otherwise, random processes will always be assumed to have zero mean so that the autocovariance and autocorrelation sequences may be used interchangeably. This assumption results in no loss of generality, since for any process $x(n)$ that hasa nonzero-mean, a zero-mean process $y(n)$ may always be formed by subtracting the mean from $x(n)$ as follows

$$y(n) = x(n) - m_x(n)$$

As in the case of random variables, the autocorrelation and autocovariance functions provide information about the degree of linear dependence between two random variables. For example, if $c_x(k,l) = 0$ for $k \neq l$, then the random variables $x(k)$ and $x(l)$ are uncorrelated and knowledge of one does not help to estimate the other using a linear estimator.

The autocovariance and the autocorrelation sequences provide information about the statistical relationship between two random variables that come from the same process, for example, $x(k)$ and $x(l)$. In applications involving more than one stochastic process, it is often of interest to determine the covariance or correlation between the random variable $x(k)$ in one process and the random variable $y(l)$ in the other process. Given two stochastic processes, $x(n)$ and $y(n)$, the cross-covariance is defined as

$$c_{xy}(k,l) = E\{[x(k) - m_x(k)][y(l) - m_y(l)]^*\} \tag{2.6}$$

And the definition of cross-correlation is also given by

$$r_{xy}(k,l) = E\{x(k)y^*(l)\} \tag{2.7}$$

These two functions satisfy the relation

$$c_{xy}(k,l) = r_{xy}(k,l) - m_x(k)m_y^*(l)$$

Just as a pair of random variables are said to be uncorrelated if $c_{xy} = 0$, two random processes, $x(n)$ and $y(n)$, are said to be uncorrelated when

$$c_{xy}(k,l) = 0$$

for all k and l or if

$$r_{xy}(k,l) = m_x(k)m_y^*(l) \tag{2.8}$$

Two random processes $x(n)$ and $y(n)$ are said to be orthogonal if their cross-correlation is zero

$$r_{xy}(k,l) = 0$$

While orthogonal stochastic processes are not necessarily uncorrelated, uncorrelated zero-mean processes are orthogonal.

$$\rho_{xy}(k,l) = \frac{r_{xy}(k,l)}{\sigma_x(k)\sigma_y(l)}$$

In any practical data acquisition device, noise or measurement error are invariably introduced into the recorded data. In many applications, this noise is modeled as additive, so if $x(n)$ denotes "signal" and $\omega(n)$ denotes "noise", the recorded signal is represented as

$$y(n) = x(n) + \omega(n)$$

Often, this additive noise is assumed to have zero mean and to be uncorrelated with the signal. In this case, the autocorrelation of the measured data, $y(n)$, is the sum of the autocorrelations of $x(n)$ and $\omega(n)$. Specifically, note that

$$\begin{aligned} r_y(k,l) = E\{y(k)y^*(l)\} &= E\{[x(k) + \omega(k)][x(l) + \omega(l)]^*\} \\ &= E\{x(k)x^*(l)\} + E\{\omega(k)\omega^*(l)\} + E\{x(k)\omega^*(l)\} + E\{\omega(k)x^*(l)\} \end{aligned}$$

If $x(n)$ and $\omega(n)$ are uncorrelated, then

$$E\{x(k)\omega^*(l)\} = E\{\omega(k)x^*(l)\} = 0$$

and it follows that

$$r_y(k,l) = r_x(k,l) + r_\omega(k,l)$$

This basic result is summarized by the following properties:

Property. If two random processes $x(n)$ and $y(n)$ are uncorrelated, then the autocorrelation of the sum

$$z(n) = x(n) + y(n)$$

is equal to the sum of the autocorrelation of $x(n)$ and $y(n)$.

$$r_z(k,l) = r_x(k,l) + r_y(k,l)$$

2.3.2 Gaussian Processes

$\boldsymbol{X}$ is called a Gaussian random vector as it's a vector of n real-valued random variables $x=[x_1, x_2, \cdots x_n]^{\mathrm{T}}$ where the random variables x_i, are said to be jointly Gaussian if the joint probability density function of the n random variables x_i satisfies:

$$f_x(x) = \frac{1}{(2\pi)^{n/2} \mid C_x \mid^{1/2}} \exp\{-(x-m_x)^{\mathrm{T}} C_x^{-1}(x-m_x)\}$$

where $\boldsymbol{m}_x = [m_1, m_2, \cdots, m_n]^{\mathrm{T}}$ is a vector containing the means of x_i ;

$$m_i = E\{x_i\}$$

$\boldsymbol{C}_x$ is a symmetric positive definite matrix with elements c_{ij} that are the covariances between x_i and x_j

$$c_{ij} = E\{(x_i - m_i)(x_j - m_j)\}$$

and $|\boldsymbol{C}_x|$ is the determinant of the covariance matrix.

A discrete-time random process $x(n)$ is said to be Gaussian if every finite set of samples from $x(n)$ are jointly Gaussian. Note that once the mean vector and covariance matrix are known, the Gaussian stochastic process is completely defined. Gaussian processes are of great interest both theoretically and practically. For example, many processes found in applications are Gaussian, or approximately Gaussian due to the central limit theorem.

2.3.3 White Noise

An important and fundamental discrete-time process that will be frequently encountered in our treatment of discrete-time stochastic processes is white noise. A wide-sense stationary process $v(n)$, either real or complex, is said to be white if for all $k \neq 0$, the autocovariance function is zero.

$$c_v(k) = \sigma_v^2 \delta(k)$$

Thus, white noise is simply a sequence of uncorrelated random variables, each with a variance of σ_v^2. Since white noise is defined only in terms of its second-order moments, there is an infinite number of white noise random processes. For example, a stochastic process consisting of an uncorrelated sequence of real-valued Gaussian random variables is a white noise process and is called White Gaussian noise (WGN). For complex white noise, note that if

$$v(n) = v_1(n) + jv_2(n)$$

then

$$E\{|v(n)|^2\} = E\{|v_1(n)|^2\} + E\{|v_2(n)|^2\}$$

Thus, it is important to note that the variance of $v(n)$ is the sum of the variances of the real and imaginary components, $v_1(n)$ and $v_2(n)$, respectively.

As we will see in Section 2.5, a wide variety of different and important random processes may be generated by filtering white noise with a linear shift-invariant filter.

2.3.4 Stationary Processes

In signal processing applications, the statistics or ensemble averages of a random process are usually independent of time. For example, quantization noise typically has a constant mean and constant variance whenever the input signal is "sufficiently complex", which results from roundoff errors in a fixed-point digital signal processor. In addition, quantization noise is usually assumed to have first-and second-order probability density functions which are independent of time. These conditions are examples of "statistical time invariance" or stationarity. In this section, several different types of stationarity are defined. As we will see in Section 2.3.5, the stationarity assumption is important for estimating the ensemble mean.

If the first-order density function of a stochastic process $x(n)$, is independent of time, i. e.

$$f_{x(n)}(\alpha) = f_{x(n+k)}(\alpha)$$

for all k, then the process is said to be first-order stationary. For a first-order stationary process, the first-order statistics will be independent of time. For example, the mean of the process will be constant

$$m_x(n) = m_x$$

and the same is true for the variance, $\sigma_x^2(n) = \sigma_x^2$. Similarly, a process is said to be second-order stationary if the second-order joint density function $f_{x(n1),x(n2)}(\alpha_1, \alpha_2)$ depends only on the difference, $n_2 - n_1$, not on the individual times n_1 and n_2. Equivalently, a process $x(n)$ is second-order stationary if k the processes $x(n)$ and $x(n+k)$ have the same second-order joint density function for any k.

$$f_{x(n_1),x(n_2)}(\alpha_1, \alpha_2) = f_{x(n_1+k),x(n_2+k)}(\alpha_1, \alpha_2)$$

If a process is second-order stationary, then it will also be first-order stationary. In addition, second-order stationary processes have second-order statistics that are invariant to a time shift of the process.

Therefore, the correlation between the random variables $x(k)$ and $x(l)$ depends only on the difference, $k-l$, separating the two random variables in time

$$r_x(k,l) = r_x(k-l,0)$$

Continuing to higher-order joint density functions, a process is said to be stationary of order L if the processes $x(n)$ and $x(n+k)$ have the same Lth-order joint density functions. Finally, a process that is stationary for all orders $L > 0$ is said to be stationary in the strict sense.

Since this book is mainly concerned with the mean and autocorrelation of processes rather than probability density functions, we will focus on another form of stationarity known as wide-sense stationarity (WSS), which is defined as follows.

Wide-Sense Stationarity. A random process $x(n)$ is said to be wide-sense stationary if the following three conditions are statisfied:

1. The mean of the process is a constant, $m_x(n) = m_x$.
2. The autocorrelation $r_x(k,l)$ depends only on the difference, $k - l$.
3. The variance of the process is finite, $c_x(0) < \infty$.

Wide-sense stationarity is a weaker constraint than second-order stationarity.

However, in the case of a Gaussian process, wide-sense stationarity is equivalent to strict-sense stationarity.

In the case of two or more processes, similar definitions exist for joint stationarity. For example, two processes $x(n)$ and $y(n)$ are said to be jointly wide-sense stationary if $x(n)$ and $y(n)$ are wide-sense stationary and if the cross-correlation $r_{xy}(k,l)$ depends only on the difference, $k-l$,

$$r_{xy}(k,l)=r_{xy}(k+n,l+n)$$

Similarly, for the joint WSS process, we will write the cross-correlation as a function only of the lag, $k-l$, as follows

$$r_{xy}(k-l)=E\{x(k)y^*(l)\}$$

The autocorrelation sequence of the WSS process has many useful and important properties, some of which are described below.

Property 1-Symmetry. The autocorrelation sequence of a WSS random process is a conjugate symmetric function of k,

$$r_x(k)=r_x^*(-k)$$

For a real process, the autocorrelation sequence is symmetric

$$r_x(k)=r_x(-k)$$

This property follows directly from the definition of the autocorrelation function

$$r_x(k)=E\{x(n+k)x^*(n)\}=E\{x^*(n)x(n+k)\}=r_x^*(-k)$$

The next property relates the value of the autocorrelation sequence at lag $k=0$ to the mean-square value of the process.

Property 2-Mean-square value. The autocorrelation sequence of a WSS process at lag $k=0$ is equal to the mean-square value of the process

$$r_x(0)=E\{|x(n)|^2\}\geqslant 0$$

As with property 1, property 2 comes directly from the definition of the autocorrelation sequence. The next property places an upper bound on the the autocorrelation sequence in terms of its value at lag $k=0$.

Property 3-Maximum value. The magnitude of the autocorrelation sequence of a WSS random process at lag k is upper bounded by its value at lag $k=0$,

$$r_x(0)\geqslant |r_x(k)|$$

This property may be established as follows

$$E\,[x(n)\ \pm x(n+m)]^2\geqslant 0$$

$$E[x^2(n)\ \pm 2x(n)x(n+k)+x^2(n+k)]\geqslant 0$$

For the WSS $x(n)$, we can get

$$E[x^2(n)]=E[x^2(n+k)]=r_x(0)$$

Substituting this into the above inequality yields

$$2r_x(0)\ \pm 2r_x(k)\geqslant 0$$

then the result follows

$$r_x(0)\geqslant |r_x(k)|$$

2. 3. 5 Ergodicity

The mean and autocorrelation of a stochastic process are examples of ensemble averages, which describe the statistical average of the process over all possible discrete-time signals. Although these ensemble averages are often needed in problems such as signal modeling, optimal filtering, and spectral estimation, they are usually not known as a priori. Therefore, being able to estimate these averages from realization of discrete-time stochastic processes becomes an important issue. In this section, we consider the estimation of the mean and autocorrelation of stochastic processes and present some conditions for which it is possible to estimate these averages using appropriate time average.

Let us first consider estimating the mean $m_x(n)$ of a random process $m_x(n)\ x(n)$. If a large number of sample realizations of the process were available, e. g. , $x_i(n), i = 1, \cdots, L$, then an average of the form

$$\hat{m}_x(n) = \frac{1}{L}\sum_{i=1}^{L} x_i(n)$$

can be used to estimate the mean $m_x(n)$. However, in most conditions, such a collection of sample realizations is usually not available, so it is necessary to consider methods for estimating the ensemble average from a single realization of the process. Given that there is only one realization of $x(n)$, we can consider estimating the ensemble average $E\{x(n)\}$ using a sample mean that is taken as the average of $x(n)$ over time as follows.

$$\hat{m}_x(N) = \frac{1}{N}\sum_{n=0}^{N-1} x(n)$$

However, imposing some constraints on the process is necessary in order for this to be a meaningful estimator for the mean.

We now consider the convergence of the sample mean of a WSS process $x(n)$ to the ensemble mean m_x. However, note that since the sample mean is the mean of a random variable, $x(0), \cdots, x(N)$, then $\hat{m}_x(N)$ is also a random variable. In fact, viewed as a sequence indexed by N, the sample mean is a sequence of random variables. Therefore, when discussing the convergence of the sample mean, it is necessary to consider convergence within a statistical framework. Although there are many different ways to formulate conditions for the convergence of a sequence of random variables, the condition of interest here is mean-square convergence, i. e.

$$\lim_{N\to\infty} E\{|\hat{m}_x(N) - m_x|^2\} = 0 \tag{2.9}$$

If equation (2. 9) is satisfied then $x(n)$ is said to be ergodic in the mean.

Definition. If the sample mean $m_x(n)$ of a wide-sense stationary process converges to m_x in the mean-square sense, then the process is said to be ergodic in the mean and we write

$$\lim_{N\to\infty} m_x(N) = m_x$$

1. The sample mean be asymptotically unbiased,

$$\lim_{N\to\infty} \hat{m}_x(N) = m_x \tag{2.10}$$

2. The variance of the estimate go to zero as $N \to \infty$

$$\lim_{N\to\infty} \mathrm{Var}\{\hat{m}_x(N)\} = 0 \tag{2.11}$$

From the definition of the sample mean it follows easily that the sample mean is unbiased for any wide-sense stationary process,

$$E\{\hat{m}_x(N)\} = \frac{1}{N}\sum_{n=0}^{N-1} E\{x(n)\} = m_x$$

However, in order to lead the variance going to zero, some constraints must be placed on the process $x(n)$. Evaluating the variance of $\hat{m}_x(N)$ we have

$$\begin{aligned}\operatorname{Var}\{\hat{m}_x(N)\} &= E\{|\hat{m}_x(N) - m|^2\} = E\left\{\left|\frac{1}{N}\sum_{n=0}^{N-1}[x(n) - m_x]\right|^2\right\} \\ &= \frac{1}{N^2}\sum_{n=0}^{N-1}\sum_{m=0}^{N-1} E\{[x(m) - m_x][x(n) - m_x]^*\} \\ &= \frac{1}{N^2}\sum_{n=0}^{N-1}\sum_{m=0}^{N-1} c_x(m - n) \qquad (2.12)\end{aligned}$$

where $c_x(m-n)$ is the autocovariance of $x(n)$. Grouping together common terms we may write the variance as

$$\begin{aligned}\operatorname{Var}\{\hat{m}_x(N)\} &= \frac{1}{N^2}\sum_{n=0}^{N-1}\sum_{m=0}^{N-1} c_x(m - n) = \frac{1}{N^2}\sum_{k=-N+1}^{N-1}(N - |k|)c_x(k) \\ &= \frac{1}{N}\sum_{k=-N+1}^{N-1}\left(1 - \frac{|k|}{N}\right)c_x(k) \qquad (2.13)\end{aligned}$$

Therefore, $x(n)$ will be ergodic in the mean if and only if

$$\lim_{N\to\infty}\frac{1}{N}\sum_{k=-N+1}^{N-1}\left(1 - \frac{|k|}{N}\right)c_x(k) = 0 \qquad (2.14)$$

An equivalent condition that is necessary and sufficient for $x(n)$ to be ergodic in the mean is given in the following theorem.

Mean Ergodic Theorem 1. Let $x(n)$ be a WSS random process with autocovariance sequence $c_x(k)$. A necessary and sufficient condition for $x(n)$ to be ergodic in the mean is

$$\lim_{N\to\infty}\frac{1}{N}\sum_{k=-N+1}^{N-1} c_x(k) = 0 \qquad (2.15)$$

Equation (2.15) places a necessary and sufficient constraint on the asymptotic decay of the autocorrelation sequence. A sufficient condition that is much easier to apply is given in the following theorem.

Mean Ergodic Theorem 2. Let $x(n)$ be a WSS random process with autocovariance sequence $c_x(k)$. Sufficient conditions for $x(n)$ to be ergodic in the mean are that $c_x(0) < \infty$ and

$$\lim_{k\to\infty} c_x(k) = 0 \qquad (2.16)$$

Thus, according to equation (2.16) a WSS process will be ergodic in the mean if it is asymptotically uncorrelated.

The mean ergodic theorems may be generalized to the estimation of other ensemble averages. For example, consider the estimation of the autocorrelation sequence

$$r_x(k) = E\{x(n)x^*(n - k)\} \qquad (2.17)$$

from a single realization of a process $x(n)$. Since, for each k, the autocorrelation is the expected value of the process

$$y_k(n) = x(n)x^*(n - k) \qquad (2.18)$$

we may estimate the autocorrelation from the sample mean of $y_k(n)$ as follows:

$$\hat{r}_x(k,N) = \frac{1}{N}\sum_{n=0}^{N-1} y_k(n) = \frac{1}{N}\sum_{n=0}^{N-1} x(n)x^*(n-k) \tag{2.19}$$

If $\hat{r}_x(k,N)$ converges in the mean-square sense to $r_x(k)$ as $N \rightarrow \infty$

$$\lim_{N\to\infty} E\{\,|\hat{r}_x(k,N) - r_x(k)|^2\} = 0$$

then the process is said to be autocorrelation ergodic and we write

$$\lim_{N\to\infty} \hat{r}_x(k,N) = r_x(k) \tag{2.20}$$

Since $\hat{r}_x(k,N)$ is the sample mean of $y_k(n)$, $y_k(n)$ it follows that $x(n)$ will be autocorrelated ergodic if $y_k(n)$ is ergodic in the mean. Applying the first-mean ergodic theorem to the sample mean of $y_k(n)$ places a constraint on the covariance of $y_k(n)$, which is equivalent to placing a constraint on the fourth-order moments of $x(n)$. Another related result applied to the Gaussian process is as follows.

Autocorrelation Ergodic Theorem. A necessary and sufficient condition for a wide-sense stationary Gaussian process with covariance $c_x(k)$ to be autocorrelation ergodic is

$$\lim_{N\to\infty} \frac{1}{N}\sum_{k=0}^{N-1} c_x^2(k) = 0 \tag{2.21}$$

Obviously, in most applications, it is not practical to determine whether a given process is ergodic. Therefore, whenever the solution for a problem requires knowledge of the mean, the autocorrelation, or some other ensemble average, it is common to assume that the process is ergodic and to use the time average to estimate these ensemble averages. The appropriateness of such an assumption will be determined by the performance of the algorithm using these estimates.

2.4 Frequency-Domain Statistical Characteristics

Just as Fourier analysis is an important tool for describing and analyzing deterministic discrete-time signals, it also plays an important role in the study of stochastic processes. However, since a stochastic process is an ensemble of discrete-time signals, we cannot compute the Fourier transform of the process itself. However, as we will see below, if we express the Fourier transform in terms of the ensemble average, it is possible to develop a frequency domain representation of the process.

Recall that the autocorrelation sequence of a WSS process provides a time-domain description of the second-order moments of the process. Since $r_k(x)$ is a deterministic sequence, we can compute its discrete-time Fourier transform.

$$P_x(\mathrm{e}^{\mathrm{j}\omega}) = \sum_{k=-\infty}^{\infty} r_k(x)\mathrm{e}^{-\mathrm{j}k\omega} \tag{2.22}$$

This is referred to as the power spectrum or power spectral density of the process. Given the power spectrum, the autocorrelation sequence can be determined by the inverse discrete-time Fourier transform of $P_x(\mathrm{e}^{\mathrm{j}\omega})$.

$$r_x(k) = \frac{1}{2\pi}\int_{-\pi}^{\pi} P_x(\mathrm{e}^{\mathrm{j}\omega})\mathrm{e}^{\mathrm{j}k\omega}d\omega \tag{2.23}$$

Thus, the power spectrum provides a frequency domain description of the second-order moment of the process. In some cases it may be more convenient to use the z-transform instead of the discrete-time Fourier transform, in which case

$$P_x(z) = \sum_{k=-\infty}^{\infty} r_k(x) z^{-k} \tag{2.24}$$

will also be referred to as the power spectrum of $x(n)$.

Just like the autocorrelation sequence, some properties of the power spectrum will be useful. First, since the autocorrelation of a WSS random process is conjugate symmetric, the power spectrum will be a real-valued function of ω. In the case of a real-valued stochastic process, the autocorrelation sequence is real and even, which means that the power spectrum is real and even. Therefore, we have the following symmetry properties of the power spectrum.

Property 1-Symmetry, The power spectrum of a WSS random process $x(n)$ is real-valued, $P_x(e^{j\omega}) = P_x^*(e^{j\omega})$, and $P_x(z)$ satisfies the symmetry condition

$$P_x(z) = P_x^*(1/z^*)$$

In addition, if $x(n)$ is real then the power spectrum is even, $P_x(e^{j\omega}) = P_x(e^{-j\omega})$ which implies that

$$P_x(z) = P_x^*(z^*)$$

In addition to being real-valued, the power spectrum is nonnegative.

Property 2–Positivity. The power spectrum of a WSS random process is nonnegative

$$P_x(e^{j\omega}) \geqslant 0$$

This property follows from the constraint that the autocorrelation matrix is nonnegative definite and will be established in the next section. Finally, a property that relates the average power in a random process to the power spectrum is as follows.

Property 3-Total Power. The power in a zero mean WSS random process is proportional to the area under the power spectral density curve

$$E\{|x(n)|^2\} = \frac{1}{2\pi}\int_{-\pi}^{\pi} P_x(e^{j\omega})\, d\omega$$

This property follows from equation (2.23) with $k = 0$ and the fact that $r_x(0) = E\{|x(n)|^2\}$

The cross-power spectral density of two zero-mean and jointly stationary stochastic processes provides a description of their statistical relations in the frequency domain and is defined as the DTFT of their cross-correlation.

The cross-Power Spectral Density (cross-PSD) is given as

$$R_{xy}(e^{j\omega}) = \sum_{l=-\infty}^{\infty} r_{xy}(l) e^{-j\omega l} \tag{2.25}$$

If the sequences $r_x(l)$ and $r_{xy}(l)$ are absolutely summable within a certain ring of the complex z plane, we can get their z -transforms

$$R_x(z) = \sum_{l=-\infty}^{\infty} r_x(l) z^{-l} \tag{2.26}$$

$$R_{xy}(z) = \sum_{l=-\infty}^{\infty} r_{xy}(l) z^{-l} \tag{2.27}$$

which are also known as the complex spectral density and complex cross-spectral density functions.

In addition to providing a frequency domain representation of the second-order moment, the power spectrum may also be related to the ensemble average of the squared Fourier magnitude, $|X(e^{j\omega})|^2$. In particular, consider

$$P_N(e^{j\omega}) = \frac{1}{2N+1}\left|\sum_{n=-N}^{N} x(n)e^{-jn\omega}\right|^2$$

$$= \frac{1}{2N+1}\sum_{n=-N}^{N}\sum_{m=-N}^{N} x(n)x^*(m)e^{-j(n-m)\omega} \tag{2.28}$$

which is proportional to the squared magnitude of the discrete-time Fourier transform of $2N+1$ samples of a given realization of the random process. For each frequency ω, since $P_N(e^{j\omega})$ is a random variable, taking the expected value it follows that

$$E\{P_N(e^{j\omega})\} = \frac{1}{2N+1}\sum_{n=-N}^{N}\sum_{m=-N}^{N} r_x(n-m)e^{-j(n-m)\omega} \tag{2.29}$$

With the substitution $k = n - m$ after some rearrangement of terms, Eq. (2.29) becomes

$$E\{P_N(e^{j\omega})\} = \frac{1}{2N+1}\sum_{k=-2N}^{N}(2N+1-|k|)r_x(k)e^{-jk\omega}$$

$$= \sum_{k=-2N}^{N}\left(1-\frac{|k|}{2N+1}\right)r_x(k)e^{-jk\omega} \tag{2.30}$$

with the substitution $k=n-m$.

Assuming that the autocorrelation sequence decays to zero fast enough so that

$$\sum_{k=-\infty}^{\infty}|k|r_x(k) < \infty$$

Take the limit of equation (2.30) by letting $N\to\infty$:

$$\lim_{N\to\infty} E\{P_N(e^{j\omega})\} = \sum_{k=-\infty}^{\infty} r_x(k)e^{-jk\omega} = P_x(e^{j\omega}) \tag{2.31}$$

Therefore, combining equations (2.28) and (2.31) we have

$$P_x(e^{j\omega}) = \lim_{N\to\infty}\frac{1}{2N+1}E\left\{\left|\sum_{n=-N}^{N} x(n)e^{-jn\omega}\right|^2\right\} \tag{2.32}$$

Thus, the power spectrum may be viewed as the expected value of $P_N(e^{j\omega})$ in the limit as $N\to\infty$

The summary of time and frenquency domain properties of stationary random are as follows:

Definitions

Mean value	$m_x = E\{x(n)\}$
Autocorrelation	$r_x(l) = E\{x(n)x^*(n-l)\}$
Autocovariance	$c_x(l) = E[x(n)-\mu_x][x(n-l)-\mu_x]^*$
Cross-correlation	$r_{xy}(l) = E\{x(n)y^*(n-l)\}$
Cross-covariance	$c_{xy}(l) = E\{[x(n)-\mu_x][y(n-l)-\mu_y]^*\}$
Power spectral density	$R_x(e^{j\omega}) = \sum_{l=-\infty}^{\infty} r_x(l)e^{-j\omega l}$

Continued

Cross-power spectral density	$R_{xy}(e^{j\omega}) = \sum_{l=-\infty}^{\infty} r_{xy}(l) e^{-j\omega l}$
Magnitude square coherence	$\lvert G_{xy}(e^{j\omega}) \rvert^2 = \lvert R_{xy}(e^{j\omega}) \rvert^2 / [R_x(e^{j\omega}) R_y(e^{j\omega})]$
Interrelations	
$\gamma_x(l) = r_x(l) - \lvert \mu_x \rvert 2$	
$\gamma_{xy}(l) = r_{xy}(l) - \mu_x \mu_y^*$	
Properties	
Autocorrelation:	**Auto-PSD**
$r_x(l)$ is non-negative definite	$R_x(e^{j\omega}) \geqslant 0$ and real
$r_x(l) = r_x^*(-l)$	$R_x(e^{j\omega}) = R_x(e^{-j\omega})$ [real $x(n)$]
$\lvert r_x(l) \rvert \leqslant r_x(0)$	$R_x(z) = R_x^*(1/z^*)$
$\lvert \rho_x(l) \rvert \leqslant 1$	$R_x(z) = R_x(z^{-1})$ [real $x(n)$]
Cross-correlation	Cross-PSD
$r_{xy}(l) = r_{yx}^*(-l)$	
$\lvert r_{xy}(l) \rvert \leqslant [r_x(0) r_y(0)]^{1/2} \leqslant \frac{1}{2}[r_x(0) + r_y(0)]$	$R_{xy}(z) = R_{yx}^*(1/z^*)$ $0 \leqslant \lvert G_{xy}(e^{j\omega}) \rvert \leqslant 1$

2.5 Filtering Random Processes

Linear shift-invariant filters are often used to perform a variety of different signal processing tasks, ranging from signal detection and estimation to deconvolution, signal representation and synthesis. Since the inputs to these filters are often random processes, it is important to be able to determine how the statistics of these processes change as a result of filtering. In this section, we derive the relationship between the mean and autocorrelation of the input processes to the mean and autocorrelation of the output processes.

Let $x(n)$ be a WSS stochastic process with mean m_x and autocorrelation $r_x(k)$. If $x(n)$ is filtered with a stable linear shift-invariant filter having unit impulse response $h(n)$, then the output $y(n)$ is a stochastic process whose relation to $x(n)$ is the convolution sum

$$y(n) = x(n) * h(n) = \sum_{k=-\infty}^{\infty} h(k) x(n-k) \tag{2.33}$$

The mean of $y(n)$ may be found by taking the expected value of equation (2.9) as follows

$$E\{y(n)\} = E\{\sum_{k=-\infty}^{\infty} h(k) x(n-k)\} = \sum_{k=-\infty}^{\infty} h(k) E\{x(n-k)\}$$

$$= m_x \sum_{k=-\infty}^{\infty} h(k) = m_x H(e^{j0}) \tag{2.34}$$

Thus, the mean of $y(n)$ is constant and it is related to the mean of $x(n)$ by a scale factor that is equal to the frequency response of the filter at $\omega = 0$.

We may also use equation (2.9) to relate the autocorrelation of $y(n)$ to the autocorrelation of $x(n)$. This is done by first computing the cross-correlation between $x(n)$ and $y(n)$ as follows,

$$r_{yx}(n+k,n) = E\{y(n+k)x^*(k)\} = E\{\sum_{l=-\infty}^{\infty} h(l)x(n+k-l)x^*(k)\}$$

$$= \sum_{l=-\infty}^{\infty} h(l)E\{x(n+k-l)x^*(n)\}$$

$$= \sum_{l=-\infty}^{\infty} h(l)r_x(k-l) \tag{2.35}$$

Thus, for a wide-sense stationary process $x(n)$, the crross-correlation $r_{yx}(n+k,n)$ depends only on the difference between $n+k$ and n and equation (2.11) may be written as

$$r_{yx}(k) = r_x(k) * h(k) \tag{2.36}$$

The autocorrelation of $y(n)$ may now be determined as follows,

$$r_y(n+k,n) = E\{y(n+k)y^*(n)\} = \sum_{l=-\infty}^{\infty} h(n-l)r_{yx}(n+k-l) \tag{2.37}$$

Chaging the index of summation by setting $m = n - l$ we have

$$r_y(n+k,n) = \sum_{l=-\infty}^{\infty} h^*(m)r_{yx}(m+k) = r_{yx}(k) * h^*(-k) \tag{2.38}$$

Therefore, the autocorrelation sequence $r_y(n+k,n)$ depends only on k, thus we have

$$r_y(k) = r_{yx}(k) * h^*(-k) \tag{2.39}$$

Combining equations (2.36) and (2.39) we have

$$r_y(k) = r_x(k) * h(k) * h^*(-k) \tag{2.40}$$

These relationships are illustrated in Figure 2.2. Thus, it follows from equation (2.34) and equation (2.40) that if $x(n)$ is WSS, then $y(n)$ will be WSS provided $\sigma_y^2 < \infty$ (this condition requires that the filter be stable). In addition, it follows from equation (2.35) that $x(n)$ and $y(n)$ will be jointly WSS.

Another interpretation of equation (2.40) is as follows. Defining $r_h(k)$ to be the (deterministic) autocorrelation of the unit impulse response, $h(n)$:

$$r_h(k) = h(k) * h^*(-k) = \sum_{n=-\infty}^{\infty} h(n)h^*(n+k)$$

we see that $r_y(k)$ is the convolution of the autocorrelation of the input process with the deterministic autocorrelation of the filter

$$r_y(k) = r_x(k) * r_h(k) \tag{2.41}$$

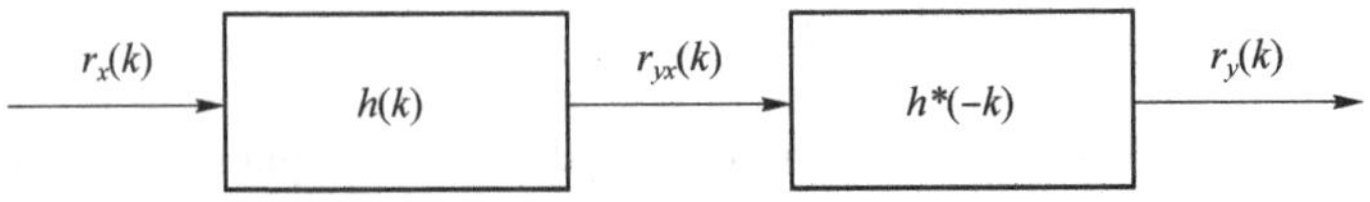

Figure 2.2 Computing the autocorrelation sequence of a filtered process $y(n)=x(n) * h(n)$

For the variance of the output process $y(n)$, we know that

$$\sigma_y^2 = r_y(0) = \sum_{l=-\infty}^{\infty} \sum_{m=-\infty}^{\infty} h(l) r_x(m-l) h^*(m)$$

Having found the relationship between the autocorrelation of $x(n)$ and the autocorrelation of $y(n)$, we may relate the power spectrum of $y(n)$ to that of $x(n)$ as follows:

$$P_y(e^{j\omega}) = P_x(e^{j\omega}) \mid H(e^{j\omega}) \mid^2 \tag{2.42}$$

Therefore, if a WSS process is filtered with a linear shift-invariant filter, then the power spectrum of the input signal is multiplied by the squared magnitude of the frequency response of the filter. In terms of z-transforms, equation (2.42) becomes

$$P_y(z) = P_x(z) H(z) H^*(1/z^*) \tag{2.43}$$

If $h(n)$ is real, then $H(z) = H^*(z^*)$ and equation (2.43) becomes

$$P_y(z) = P_x(z) H(z) H(1/z) \tag{2.44}$$

Thus, if $h(n)$ is complex and the system function $H(z)$ has a pole at $z = z_0$, then the power spectrum $P_y(z)$ will have a pole at $z = z_0$ and another at the conjugate reciprocal location $z = 1/z_0^*$. Similarly, if $H(z)$ has a zero at $z = z_0$, then the power spectrum of $y(n)$ will have a zero at $z = z_0$ as well as a zero at $z = 1/z_0^*$. This is a special case of spectral factorization as discussed in the following section.

2.6 Special Types of Random Processes

In this section, we look at the characteristics and properties of some stochastic processes that we will encounter in the following chapters. We begin by looking at those processes that can be generated by filtering white noise with a linear shift-variance filter having a rational system function. These processes include autoregressive and autoregressive moving average processes. The form of the autocorrelation sequence and the power spectrum of the these processes will be of interest. For example, we will show that the autocorrelation sequences of these processes satisfy a set of equations called Yule-Walker equation, relating $r_x(k)$ to the parameters of the filter. In addition to the filtered white noise processes, we will also study harmonic processes consisting of a sum of random phase sine waves or complex exponentials. These processes are important in applications where the signal has periodic components.

2.6.1 Autoregressive Moving Average Processes

Suppose that we filter white noise $v(n)$ with a causal linear shift-invariant filter having a rational system function with p poles and q zeros

$$H(z) = \frac{B_q(z)}{A_p(z)} = \frac{\sum_{k=0}^{q} b_q(k) z^{-k}}{1 + \sum_{k=1}^{p} a_p(k) z^{-k}} \tag{2.45}$$

Assuming that the filter is stable, the output process $x(n)$ will be wide-sense stationary and, with $p_v(z) = \sigma_v^2$, the power spectrum will be

$$P_x(z) = \sigma_v^2 \frac{B_q(z)B_q^*(1/z^*)}{A_p(z)A_p^*(1/z^*)} \tag{2.46}$$

A process having a power spectrum of this form is known as an autoregressive moving average process of order (p,q) and is referred to as an ARMA (p,q) process.

Since $v(n)$ and $x(n)$ are related by the linear constant coefficient difference equation

$$x(n) + \sum_{l=1}^{q} a_q(l)x(n-l) = \sum_{l=0}^{q} b_q(l)v(n-l) \tag{2.47}$$

Since $v(n)$ is WSS, it follows that $v(n)$ and $x(n)$ are jointly WSS and

$$E\{v(n-l)x^*(n-k)\} = r_{vx}(k-l) \tag{2.48}$$

Therefore, we have

$$r_x(k) + \sum_{l=1}^{q} a_q(l)r_x(k-l) = \sum_{l=0}^{q} b_q(l)r_{vx}(k-l) \tag{2.49}$$

which is a difference equation of the same form as equation (2.33). In its present form, this difference equation is not very useful. However, by writing the cross-correlation $r_{vx}(k)$ in terms of autocorrelation $r_x(k)$ and the unit impulse response of the filter, we may derive a set of equations relating the autocorrelation of $x(n)$ to the filter. The cross-correlation may be written as

$$E\{v(n-l)x^*(n-k)\} = \sigma_v^2 h^*(l-k) \tag{2.50}$$

2.6.2 Autoregressive Processes

A special type of ARMA (p,q) process results when $q=0$ and $B_q(z)=b(0)$. In this case $x(n)$ is generated by filtering white noise with an all-pole filter of the form

$$H(z) = \frac{b(0)}{1 + \sum_{k=1}^{p} a_p(k)z^{-k}} \tag{2.51}$$

An ARMA $(p,0)$ process is called an autoregressive process of order p and will be referred to as an AR (p) process, the power spectrum of $x(n)$ is

$$P_x(z) = \sigma_v^2 \frac{|b(0)|^2}{A_p(z)A_p^*(1/z^*)} \tag{2.52}$$

Thus, the power spectrum of an AR (p) process contains $2p$ poles and no zeros (except those located at $z=0$ and $z=\infty$). With $c_0(0) = b(0)h^*(0) = |b(0)|^2$, these equations are

$$r_x(k) + \sum_{l=1}^{q} a_q(l)r_x(k-l) = \sigma_v^2 |b(0)|^2\delta(k) \tag{2.53}$$

Finally, we have $r_x(k) = \dfrac{b^2(0)}{1-a^2(1)}[-a(1)]^{|k|}$

2.6.3 Harmonic Processes

Harmonic processes provide useful representations for random processes that arise in applications such as array processing, when the signals contain periodic components. An example of a wide-sense stationary harmonic process is the random phase sinusoid

$$x(n) = A\sin(n\omega_0 + \phi)$$

where A and ω_0 are constants and ϕ is a random variable that is uniformly distributed between $-\pi$ and π. The autocorrelation sequence of $x(n)$ is periodic with frequency ω_0

$$r_x(n) = \frac{1}{2}A^2\cos(k\omega_0) \tag{2.54}$$

Therefore, the power spectrum is

$$P_x(e^{j\omega}) = \frac{\pi}{2}A^2[u_0(\omega - \omega_0) + u_0(\omega + \omega_0)]$$

where $u_0(\omega - \omega_0)$ is an impulse at frequency ω_0 and $u_0(\omega + \omega_0)$ is an impulse at frequecy $-\omega_0$. If the amplitude is also a random variable that is uncorrelated with ϕ, then the autocorrelation sequence will be

$$r_x(n) = \frac{1}{2}E\{A^2\}\cos(k\omega_0)$$

Higher-order harmonic processes may be formed from a sum of random phase sinusoids as follows:

$$x(n) = \sum_{l=1}^{L} A_l \sin(n\omega_l + \phi_l)$$

Assuming the random variables ϕ_l and A_l are uncorrelated, the autocorrelation sequence is

$$r_x(n) = \sum_{l=1}^{L} \frac{1}{2}E\{A_l^2\}\cos(k\omega_l)$$

and the power spectrum is

$$P_x(e^{j\omega}) = \sum_{l=1}^{L} \frac{\pi}{2}E\{A^2\}[u_0(\omega - \omega_l) + u_0(\omega + \omega_l)]$$

In the case of complex signals, a first order harmonic process has the form

$$x(n) = |A|e^{j(n\omega_0+\phi)}$$

Then the autocorrelation sequence is a complex exponential with frequency ω_0

$$r_x(k) = E\{|A|^2\}e^{jk\omega_0}r_x(k)$$

and the power spectrum is

$$P_x(e^{j\omega}) = 2\pi E\{|A|^2\}u_0(\omega - \omega_0)$$

With a sum of Luncorrelated harmonic processes

$$x(n) = \sum_{l=1}^{L} A_l e^{[j(n\omega_l+\varphi_l)]}$$

the autocorrelation sequence is

$$r_x(k) = \sum_{l=1}^{L} E\{| A_l |^2\}e^{jk\omega_l}$$

and the power spectrum is

$$P_x(e^{j\omega}) = \sum_{l=1}^{L} 2\pi E\{| A |^2\}u_0(\omega - \omega_l)$$

2.7 Summary

In this chapter, we introduced the basic theory of discrete-time stochastic processes. A complete description of a stochastic process usually requires a statement of the joint distribution or density functions, however, in this book our main concern is its ensemble averages. Therefore, the mean, the variance, and the autocorrelation of discrete-time stochastic processes were firstly defined. Next, we determined the concept of stationarity because it is useful to reduce the computational complexity in many applications. Assuming time invariance on the first-and second-order moments, we defined a wide-sense stationary (WSS) process in which the mean is a constant and correlation between random variables at two distinct times is a function of time difference or lag. The rest of the chapter was decicated to the analysis of WSS processes.

In practice, we typically consider a stochastic process as a single sample function and estimating the first-and the second-order moments is necessary. This needs the notion of ergodicity that provides a framework for the computation of statistical averages using time averages over a single realization. Although this framework requires theoretical results using mean square convergence, we provided a simple approach of using appropriate time averages. Then we also discussed two useful and interesting average ergodicity theorems and an autocorrelation ergodicity theorem for Gaussian processes. In this book, we won't focuse on whether the conditions of the ergodic theorem are satisfied when an ensemble average needs to be estimated, since these theorems cannot be applied without some prior knowledge of higher-order statistics. Instead we will assume that, the ensemble average can be estimated by simply taking an appropriate time average. In practice, the ultimate justification for this assumption is the excellent performance of the algorithms that use these estimates.

As another important statistical characteristic, the power spectrum was introduced in this chapter, which is the discrete-time Fourier transform of a wide-sense stationary process. It is a significant tool for describing, modeling, and analyzing wide-sense stationary processes. Next, we explored the process of filtering a wide-sense stationary process with a discrete-time filter. It is particularly important how the mean, autocorrelation and power spectrum of the input process are changed by the filter. Finally, we concluded this chapter with a list of some important types of processes that will be encountered throughout the book. These processes include those that result from filtering white noise with a linear shift-invariant filter, such as autoregressive and autoregressive moving average processes. As a complement, we also described harmonic processes, which consist of a sum of random phase sinusoids or complex exponentials. These topics will be useful in many subsequent chapters.

Exercises

1. Let $x(n)$ be a stationary random process with zero mean and autocorrelation $r_x(k)$. We form the process, $y(n)$, as follows

$$y(n) = x(n) + f(n)$$

where $f(n)$ is a known deterministic sequence. Find the mean $m_y(n)$ and the autocorrelation $r_y(k,l)$ of the process $y(n)$.

Solution:

The mean of the process is

$$m_y(n) = E\{y(n)\} = E\{x(n)\} + f(n) = f(n)$$

and the autocorrelation is

$$r_y(k,l) = E\{y(k)y(l)\} = E\{[x(k) + f(k)][x(l) + f(l)]\}$$
$$= E\{x(k)x(l)\} + f(k)f(l) = r_x(k,l) + f(k)f(l)$$

2. A discrete-time random process $x(n)$ is generated as follows

$$x(n) = \sum_{k=1}^{p} a(k)x(n-k) + w(n)$$

where $\omega(n)$ is a white noise process with variance σ_ω^2. Another process, $z(n)$, is formed by adding noise to $x(n)$

$$z(n) = x(n) + v(n)$$

where $v(n)$ is white noise with a variance of σ_v^2 that is uncorrelated with $\omega(n)$.

(a) Find the power spectrum of $x(n)$;

(b) Find the power spectrum of $z(n)$.

Solution:

(a) Since $x(n)$ is the output of an all-pole filter driven by white noise, $x(n)$ is an $AR(p)$ process with a power spectrum

$$P_x(e^{j\omega}) = \frac{\sigma_\omega^2}{|A(e^{j\omega})|^2}$$

where

$$A(e^{j\omega}) = 1 - \sum_{k=1}^{p} a(k)e^{-jk\omega}$$

(b) The process $z(n)$ is a sum of two random process

$$z(n) = x(n) + v(n)$$

Since $x(n)$ is a linear combination of values of $\omega(n)$,

$$x(n) = \sum_{k=-\infty}^{n} h(k)\omega(n-k)$$

where $h(n)$ is the unit impulse response of the filter generating $x(n)$, and since $v(n)$ is uncorrelated with $\omega(n)$, then $v(n)$ is uncorrelated with $x(n)$, and we have

$$r_z(k) = r_x(k) + r_v(k)$$

therefore,

$$P_z(e^{j\omega}) = P_x(e^{j\omega}) + P_v(e^{j\omega})$$

and

$$P_z(e^{j\omega}) = \frac{\sigma_\omega^2}{|A(e^{j\omega})|^2} + \sigma_v^2 = \frac{\sigma_\omega^2 + \sigma_v^2|A(e^{j\omega})|^2}{|A(e^{j\omega})|^2}$$

3. Periodogram function is used to calculate the power spectrum (direct power spectrum estimation). Suppose a signal consists of a frequency of 100 Hz and 500 Hz, with a sampling frequency of 2 000 Hz and 1 024 sampling points, use the function periodogram to estimate its power spectrum.

Solution:

Some relevant codes are as follows and Figure 2. 3 shows the power spectrum using direct method.

```
Fs=2000;
nFFT=1024;
n=0:1/Fs:1;
x=sin(2 * pi * 100 * n)+4 * sin(2 * pi * 500 * n)+randn(size(n);
window=boxcar(length(x));
periodogram(x,window,NFFT,Fs);
```

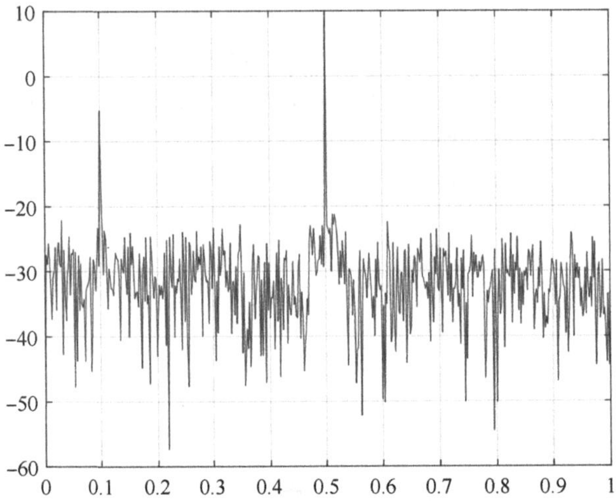

Figure 2. 3 power spectrum using direct method

4. Under the same conditions in topic 3, use an indirect method to estimate the power spectrum of this signal.

Solution:

Some relevant codes are as follows and Figure 2. 4 shows the power spectrum using inderect method.

```
Fs=2000;
nFFT=1024;
n=0:1/Fs:1;
x=sin(2 * pi * 100 * n)+4 * sin(2 * pi * 500 * n)+randn(size(n);
Cx=xcorr(x,'unbiased');
Cxk=fft(Cx,nFFT);
Pxx=abs(Cxk);
t=0:round(nFFT/2-1);
k=t * Fs/nFFT;
P=10 * log(Pxx(t+1));
plot(k,P);
```

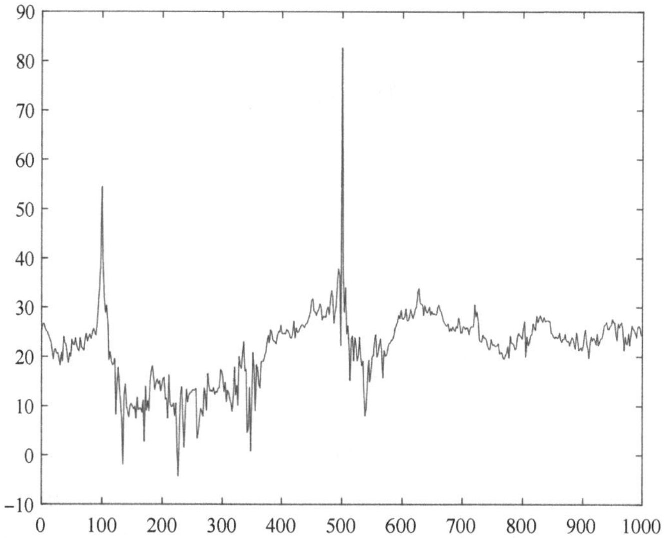

Figure 2. 4 power spectrum using indirect method

References

[1] G. J. Olsder and J. A. C. Resing, "Discrete event systems with stochastic processing times", IEEE Transactions on Automatic Control, Vol. 35, pp. 299–302, 1990.

[2] C. W. Therrien, "Discrete Random Signals and Statistical Signal Processing", Prentice-Hall, Englewood Cliffs, NJ, 1992.

[3] John G. Proakis and Charles M. Rader, "Algorithms for Statistical Signal Processing", Tsinghua University, 2003.

[4] A. V. Oppenheim and R. W. Schafer, "Discrete-Time Signal Processing", Prentice-Hall, Englewood Cliffs, NJ, 1989.

[5] B. Picinbono, "Random Signals and Noise", Prentice-Hall, Englewood Cliffs, NJ, 1993.

Chapter 3
Signal Modeling

3.1 Introduction

In this chapter, we will introduce and analyze the properties of a special class of stationary random sequences. Stationary random sequences are obtained by filtering white noise through a linear, time-invariant system. We are concerned with filters with a rational system function, i. e. , the ratio of two polynomials. The power spectral density of the output process thus obtained is also rational, and its shape is entirely determined by the coefficients of the filter.

When we want to emphasize the systems viewpoint, we will use the term pole-zero model and the term autoregressive moving average model to refer to generated random sequences. The latter term is not appropriate when the input is a harmonic process or a deterministic signal with a flat spectral envelope.

We discuss the impulse response, autocorrelation, power spectrum, partial autocorrelation, and frequency spectrum for the all-pole, all-zero, and all-pole-zero models. We use model coefficients to represent all these quantities and find ways to convert from one set of parameters to another. Low-order models are easy to analyze and help to familiarize with the behavior and properties of higher-order models, so this chapter examines low-order models in detail. Understanding the correlation and spectral properties of the signal model is important for selecting an appropriate model in practical applications.

The simplest model of a random signal is a wide-sense stationary white noise sequence with uncorrelated characteristics and a flat Power Spectral Density (PSD). If we filter the white noise with a stable LTI filter, we can obtain a random signal with almost any arbitrary non-periodic correlation structure or continuous PSD. If we wish to generate a random signal with a linear PSD by using the previous method, we need an LTI filter with a "linear" frequency response; that is, we need an oscillator. Unfortunately, such a system is not stable, and its output cannot be stationary. Fortunately, a random signal with a line PSD can be easily generated by using a harmonic process model (a linear combination of sinusoidal wave sequences with statistically independent random phases). Figure 3.1 below illustrates the filtering of white noise and "white" (flat spectral envelope) harmonic processes with an LTI filter. The signal model with mixed PSD can be obtained by combining the above two models, and this process is demonstrated by a powerful result called, the Wold decomposition.

When the LTI filter is specified as an impulse response, we have a nonparametric signal model because there is no restriction on the form of the model and the number of parameters is infinite. However, if we specify the filter with a finite order rational system function, we have a parametric signal model described by a finite number of parameters. We focus on parametric models because they are simpler to handle in practical applications. The two main topics we discuss in this chapter are (1) the derivation of the second-order moments of the AP, AZ, and PZ models for a given coefficient of the system function, and (2) the design of an AP, AZ, or PZ system that produces a random signal with a given autocorrelation sequence or PSD function. The second problem is known as the signal modeling.

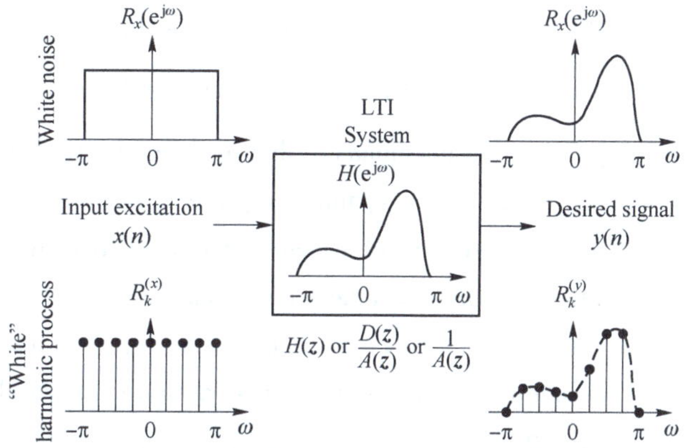

Figure 3.1 Signal models with continuous and discrete (line) power spectrum densities

3.1.1 Linear Nonparametric Signal Models

Consider a stable LTI system with impulse response $h(n)$ and input $w(n)$. The output $x(n)$ is given by the convolution summation;

$$x(n) = \sum_{k=-\infty}^{\infty} h(k)w(n-k) \tag{3.1}$$

which is known as anonrecursive system representation because the output is computed by linearly weighting samples of the input signal.

Linear random signal model. If the input $w(n)$ is a zero-mean white noise process with variance σ_w^2, autocorrelation $r_w(l) = \sigma_w^2\delta(l)$, and PSD $R_w(e^{j\omega}) = \sigma_w^2$, $-\pi < \omega < \pi$, then from the autocorrelation, complex PSD, and PSD of the output $x(n)$ are respectively given by

$$r_x(l) = \sigma_w^2 \sum_{k=-\infty}^{\infty} h(k) * h(k-l) = \sigma_w^2 r_h(l) \tag{3.2}$$

$$R_x(z) = \sigma_w^2 H(z) H^*\left(\frac{1}{z^*}\right) \tag{3.3}$$

$$R_x(e^{j\omega}) = \sigma_w^2 \, |H(e^{j\omega})|^2 = \sigma_w^2 R_h(e^{j\omega}) \tag{3.4}$$

We notice that when the input is a white noise process, the shape of the autocorrelation and the power spectrum (second-order moments) of the output signal are completely characterized by the system. We use the term system-based signal model to refer to the signal generated by a system with a white noise input. If the system is linear, we use the term linear random signal model. In the statistical literature, the resulting model is known as the general linear process model. However, we should mention that in some applications it is more appropriate to use a deterministic input with flat spectral envelope or a "white" harmonic process input.

Recursive representation. Suppose now that the inverse system $H_I(n) = 1/H(z)$ is causal and stable. If we assume, without any loss of generality, that $h(0) = 1$, then $h_I(n) = Z^{-1}\{H_I(n)\}$ has $h_I(0) = 1$. Therefore the input $w(n)$ can be obtained by

$$w(n) = x(n) + \sum_{k=1}^{\infty} h_I(k)x(n-k) \tag{3.5}$$

Solving for $x(n)$, we obtain the following recursive representation for the output signal

$$x(n) = -\sum_{k=-1}^{\infty} h_I(k)x(n-k) + w(n) \tag{3.6}$$

We use the term recursive representation to emphasize that the present value of the output is obtained by a linear combination of all past output values, plus the present value of the input. By construction the nonrecursive and recursive representations of system $h(n)$ are equivalent; that is, they produce the same output when they are excited by the same input signal.

Innovations representation. If the system $H(z)$ is minimum-phase, then both $h(n)$ and $h_I(n)$ are causal and stable. Hence, the output signal can be expressed nonrecursively by

$$x(n) = \sum_{k=0}^{\infty} h(k)x(n-k) = \sum_{k=-\infty}^{n} h(n-k)w(k) \tag{3.7}$$

or recursively by equation (3.6).

From equation (3.7) we obtain

$$x(n+1) = \sum_{k=-\infty}^{n} h(n+1-k)w(k) + w(n+1)$$

or by using equation (3.5)

$$x(n+1) = \underbrace{\sum_{k=-\infty}^{n} h(n+1-k) \sum_{j=-\infty}^{n} h_1(k-j)x(j)}_{\text{past information: linear combination of } x(n),\, x(n-1),\, \cdots} + \underbrace{w(n+1)}_{\text{new information}} \tag{3.8}$$

Careful inspection of equation (3.8) indicates that if the system generating $x(n)$ is minimum-phase, the sample $w(n+1)$ brings all the new information (*innovation*) to be carried by the sample $x(n+1)$. All other information can be predicted from the past samples $x(n), x(n+1), \cdots$ of the signal. We stress that this interpretation holds only if $H(z)$ is minimum-phase.

The system $H(z)$ generates the signal $x(n)$ by introducing dependence in the white noise input $w(n)$ and is known as the synthesis or coloring filter. In contrast, the inverse system $H_I(z)$ can be used to recover the input $w(n)$ and is known as the analysis or whitening filter. In this sense the innovations sequence and the output process are completely equivalent.

The synthesis and analysis filters are shown in Figure 3.2.

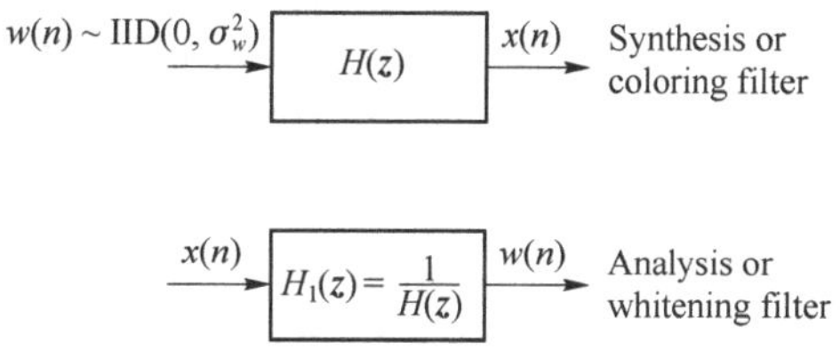

Figure 3.2 Synthesis and analysis filters used in innovations representation

Spectral factorization

Most random processes with a continuous PSD $R_x(e^{j\omega})$ can be generated by exciting a minimum-phase system $H_{\min}(z)$ with white noise. The PSD of the resulting process is given by

$$R_x(e^{j\omega}) = \sigma_w^2 \, |H_{\min}(e^{j\omega})|^2 \tag{3.9}$$

The process of obtaining $H_{\min}(z)$ from $R_x(e^{j\omega})$ or $r_x(l)$ is known as spectral factorization. If the PSD $R_x(e^{j\omega})$ satisfies the Paley-Wiener condition

$$\int_{-\pi}^{\pi} |\ln R_x(e^{j\omega})| d\omega < \infty \tag{3.10}$$

then the process $x(n)$ is called regular and its complex PSD can be factored as follows

$$R_x(z) = \sigma_w^2 H_{\min}(z) H_{\min}{}^*\left(\frac{1}{z^*}\right) \tag{3.11}$$

where

$$\sigma_w^2 = \exp\left\{\frac{1}{2\pi}\int_{-\pi}^{\pi} \ln[R_x(e^{j\omega})]d\omega\right\} \tag{3.12}$$

is the variance of the white noise input and can be interpreted as the geometric mean of $R_x(e^{j\omega})$. Consider the inverse Fourier transform of $\ln R_x(e^{j\omega})$

$$c(k) \triangleq \frac{1}{2\pi}\int_{-\pi}^{\pi} \ln[R_x(e^{j\omega})]e^{jk\omega}d\omega \tag{3.13}$$

which is a sequence known as the cepstrum of $r_x(l)$. Note that $c(0) = \sigma_w^2$. Thus in the cepstral domain, the multiplicative factors $H_{\min}(z)$ and $H_{\min}^*\left(\frac{1}{z^*}\right)$ are now additively separable due to the natural logarithm of $R_x(e^{j\omega})$. Define

$$c_+(k) \triangleq \frac{c(0)}{2} + c(k)u(k-1) \tag{3.14}$$

and

$$c_-(k) \triangleq \frac{c(0)}{2} + c(k)u(-k-1) \tag{3.15}$$

as the positive and negative-axis projections of $c(k)$, respectively, with $c(0)$ distributed equally between them. Then we obtain

$$h_{\min}(n) = \mathcal{F}^{-1}\{\exp \mathcal{F}[c_+(k)]\} \tag{3.16}$$

as the impulse response of the minimum-phase system $H_{\min}(z)$. Similarly,

$$h_{\max}(n) = \mathcal{F}^{-1}\{\exp \mathcal{F}[c_-(k)]\} \tag{3.17}$$

is the corresponding maximum-phase system. This completes the spectral factorization procedure for an arbitrary PSD $R_x(e^{j\omega})$, which in general is a complicated task. However, it is straightforward if $R_x(z)$ is a rational function.

Spectral Flatness Measure

The Spectral Flatness Measure (SFM) of a zero-mean process with PSD $R_x(e^{j\omega})$ is defined by Makhoul

$$\text{SFM}_x \triangleq \frac{\exp\left\{\frac{1}{2\pi}\int_{-\pi}^{\pi}\ln[R_x(e^{j\omega})]\,d\omega\right\}}{\frac{1}{2\pi}\int_{-\pi}^{\pi}R_x(e^{j\omega})\,d\omega} = \frac{\sigma_w^2}{\sigma_x^2} \tag{3.18}$$

where the second equality follows from equation (3.12). It describes the shape of the PSD by a single number. If $x(n)$ is a white noise process, then $R_x(e^{j\omega}) = \sigma_x^2$ and $\text{SFM}_x = 1$. More specifically, we can show that

$$0 \leqslant \text{SFM}_x \leqslant 1 \tag{3.19}$$

Observe that the numerator of equation (3.18) is the geometric mean while the denominator is the arithmetic mean of a real-valued, non-negative continuous waveform $R_x(e^{j\omega})$. Since $x(n)$ is a regular process satisfying equation (3.10), these means are always positive. Furthermore, their ratio, by definition, is never greater than unity and is equal to unity if the waveform is constant. This then proves equation (3.19). A detailed proof is given in Jayant and Noll's literatures.

When $x(n)$ is obtained by filtering the zero-mean white noise process $w(n)$ through the filter $H(z)$, then the coloring of $R_x(e^{j\omega})$ is due to $H(z)$. In this case, $R_x(e^{j\omega}) = \sigma_w^2\,|H(e^{j\omega})|^2$ from equation (3.19), and we obtain

$$\text{SFM}_x = \frac{\sigma_w^2}{\sigma_x^2} = \frac{\sigma_w^2}{\frac{1}{2\pi}\int_{-\pi}^{\pi}\sigma_w^2\,|H(e^{j\omega})|^2\,d\omega} = \frac{1}{\frac{1}{2\pi}\int_{-\pi}^{\pi}|H(e^{j\omega})|^2\,d\omega} \tag{3.20}$$

Thus SFM_x is the inverse of the filter power (or power transfer factor) if $h(0)$ is normalized to unity.

3.1.2 Mixed Processes and Wold Decomposition

An arbitrary stationary random process can be constructed to possess a continuous PSD $R_x(e^{j\omega})$ and a discrete power spectrum $R_x(k)$. Such processes are called mixed processes because the continuous PSD is due to regular processes while the discrete spectrum is due to harmonic (or almost periodic) processes. A further interpretation of mixed processes is that the first part is an unpredictable process while the second part is a predictable process (in the sense that past samples can be used to exactly determine future samples). This interpretation is due to the Wold decomposition theorem.

THEOREM 3.1 (WOLD DECOMPOSITION). A general stationary random process can be written as a sum

$$x(n) = x_r(n) + x_p(n) \tag{3.21}$$

where $x_r(n)$ is a regular process possessing a continuous spectrum and $x_p(n)$ is a predictable process possessing a discrete spectrum. Furthermore, $x_r(n)$ is orthogonal to $x_p(n)$; that is

$$E\{x_r(n_1)x_p^*(n_2)\} = 0 \quad \text{for all } n_1, n_2 \tag{3.22}$$

The proof of this theorem is very involved, but a good approach to it is given in Therrien's literature. Using equation (3.22), the correlation sequence of $x(n)$ in equation (3.21) is given by

$$r_x(l) = r_{x_r}(l) + r_{x_p}(l) \tag{3.23}$$

from which we obtain the continuous and discrete spectra. As discussed above, the regular process has an innovations representation $w(n)$ that is uncorrelated but not independent. For example, $w(n)$ can be the output of an all-pass filter driven by an independent and indentically distributed (IID) sequence.

3.2 All-Pole Models

We begin by discussing linear signal models with all-pole models, as they are the easiest to analyze and most commonly used in practical applications. We assume an all-pole model of the form

$$H(z) = \frac{b(0)}{A(z)} = \frac{b(0)}{1 + \sum_{k=1}^{p} a_p(k)z^{-k}} = \frac{b(0)}{\prod_{k=1}^{p}(1 - b_p(k)z^{-1})} \tag{3.24}$$

where $b(0)$ is the system gain and p is the order of the model. The all-pole model can be implemented using either a direct or a lattice structure.

3.2.1 Properties of All-Pole Models

In this section, we derive analytical expressions for various properties of the all-pole model, namely impulse response, autocorrelation, and spectrum. Because the results can be easily used to obtain the properties of signal models for inputs with both continuous and discrete spectra, we first determine the system-related properties $r_h(l)$ and $R_h(e^{j\omega})$.

Impulse response. The impulse response $h(n)$ can be specified by first rewriting equation (3.24) as

$$H(z) + \sum_{k=1}^{p} a_p(k)H(z)z^{-k} = b(0)$$

and then taking the inverse z-transform to get

$$h(n) + \sum_{k=1}^{p} a_p(k)h(n-k) = b(0)\delta(n) \tag{3.25}$$

If the system is causal, then

$$h(n) = -\sum_{k=1}^{p} a_p(k)h(n-k) + b(0)\delta(n) \tag{3.26}$$

If $H(z)$ has all its poles inside the unit circle, then $h(n)$ is a causal, stable sequence and the system is minimum-phase. From equation (3.26) we have

$$h(0) = b(0) \tag{3.27}$$

$$h(n) = -\sum_{k=1}^{p} a_p(k)h(n-k) \qquad n > 0 \tag{3.28}$$

and owing to causality we have

$$h(n) = 0 \qquad n < 0 \tag{3.29}$$

Therefore, the value of $h(n)$ can be recursively obtained by linear weighting summation of its previous values $h(n-1), \cdots, h(n-p)$ except for the value at $n = 0$. It can be said that can be predicted (with zero error for $n \neq 0$) from the past p values. Therefore, these coefficients $\{a_p(k)\}$ are often referred to as prediction coefficients. Note that there is a close relationship between all-pole models and linear prediction, which will be discussed in Section 3.2.2.

From equation (3.27) and equation (3.28), we can also write the inverse relationship

$$a_p(n) = -\frac{h(n)}{h(0)} - \sum_{k=1}^{n-1} a_p(k)\frac{h(n-k)}{h(0)} \qquad 1 \leqslant n \leqslant p \tag{3.30}$$

with $a_p(0) = 1$. From equation (3.30) and equation (3.27), we conclude that, if the first $p+1$ values of the impulse response $h(n)$ are given $0 \leqslant n \leqslant p$, the parameters of the all-pole filter are all fully determined.

Finally, we note that a causal $H(z)$ can be written as a one-sided infinite polynomial $H(z) = \sum_{n=0}^{\infty} h(n)z^{-n}$. This representation means that any finite-order and all-pole model can be equivalently represented by an infinite number of zeros. In general, a pole can be represented by an infinite number of zeros, and conversely, a zero can be represented by an infinite number of poles. If the poles are inside the unit circle, the corresponding zeros are also inside the unit circle, and vice versa.

Autocorrelation. The impulse response of the all-pole model has infinite duration, so its autocorrelation function involves an infinite sum, and it is not feasible to write it in closed form except for low-order models. However, the autocorrelation function follows a recursive relationship that associates autocorrelation values with model parameters. Multiplying equation (3.25) by $h^*(n-l)$ and summing over all n, we have

$$\sum_{n=-\infty}^{\infty}\sum_{k=0}^{p} a_p(k)h(n-k)h^*(n-l) = b(0)\sum_{n=-\infty}^{\infty} h^*(n-l)\delta(n) \tag{3.31}$$

where $a_p(0) = 1$. Interchanging the order of summations in the left-hand side, we obtain

$$\sum_{k=0}^{p} a_p(k)r_h(l-k) = b(0)h^*(-l) \qquad -\infty < l < \infty \tag{3.32}$$

where $r_h(l)$ is the autocorrelation of $h(n)$. Equation (3.32) is true for all l, but because $h(l) = 0$ for $l < 0$, $h(-l) = 0$ for $l > 0$, and we have

$$\sum_{k=0}^{p} a_p(k)r_h(l-k) = 0 \qquad l > 0 \tag{3.33}$$

From equation (3.27) and equation (3.32), we also have for $l = 0$,

$$\sum_{k=0}^{p} a_p(k)r_h(-k) = |b(0)|^2 \tag{3.34}$$

where we used the fact that $r_h^*(-l) = r_h(l)$. Equation (3.33) can be rewritten as

$$r_h(l) = -\sum_{k=1}^{p} a_p(k) r_h(l-k) \qquad l > 0 \tag{3.35}$$

which is a recursive relation for $r_h(l)$ in terms of past values of the autocorrelation and $\{a_p(k)\}$. Relation equation (3.35) for $r_h(l)$ is similar to relation equation (3.28) for $h(n)$, but with one important difference: equation (3.28) for $h(n)$ is true for all $n \neq 0$ while equation (3.35) for $r_h(l)$ is true only if $l > 0$; for $l < 0, r_h(l)$ obeys equation (3.32).

If we define the normalized autocorrelation coefficients as

$$\rho_h(l) = \frac{r_h(l)}{r_h(0)} \tag{3.36}$$

then we can divide equation (3.34) by $r_h(0)$ and deduce the following relation for $r_h(0)$

$$r_h(0) = \frac{|b(0)|^2}{1 + \sum_{k=1}^{p} a_p(k) \rho_h(k)} \tag{3.37}$$

which is the energy output by the all-pole filter when a single impulse is excited.

Autocorrelation in terms of poles. The complex spectrum of the AP(P) model is

$$R_h(z) = H(z) H\left(\frac{1}{z^*}\right) = |b(0)|^2 \prod_{k=1}^{p} \frac{1}{(1 - b_p(k) z^{-1})(1 - b_p(k) z^*)} \tag{3.38}$$

Therefore, the autocorrelation sequence can be obtained in terms of the poles by taking the inverse z-transform of $R_h(z)$, that is, $r_h(l) = Z^{-1}\{R_h(z)\}$. The poles p_k of the minimum-phase model $H(z)$ contribute causal terms in the partial fraction expansion, whereas the poles $1/p_k$ of the nonminimum-phase model $H(1/z^*)$ contribute noncausal terms.

Impulse train excitations. The response of an AP(P) model to a periodic impulse train with period L is periodic with the same period and is given by

$$\tilde{h}(n) + \sum_{k=1}^{p} a_p(k) \tilde{h}(n-k) = b(0) \sum_{m=-\infty}^{\infty} \delta(n + Lm)$$

$$= \begin{cases} b(0) & n + Lm = 0 \\ 0 & n + Lm \neq 0 \end{cases} \tag{3.39}$$

which shows that the prediction error is zero for the period, and $b(0)\ d_0$ at the beginning of each period. If we multiply both sides of equation (3.39) by $\tilde{h}^*(n-l)$ and sum over a period $0 \leqslant n \leqslant L-1$, we obtain

$$\tilde{r}_h(l) + \sum_{k=1}^{p} a_p(k) \tilde{r}_h(l-k) = \frac{b(0)}{L} \tilde{h}^*(-l) \ all\ l \tag{3.40}$$

where $\tilde{r}_h(l)$ is the periodic autocorrelation of $\tilde{h}(n)$. Since, in contrast to $h(n)$ in equation (3.32), $\tilde{h}(n)$ is not necessarily zero for $n < 0$, the periodic autocorrelation $\tilde{r}_h(l)$ generally no longer follows the linear prediction equation. Similar results can be obtained for harmonic process excitations.

Model parameters in terms of autocorrelation. Equation (3.32) for $l = 0, 1, \cdots, p$ comprise $p+1$ equations that relate the $p+1$ parameters of $H(z)$, namely, $b(0)$ and $\{a_p(k), 1 \leqslant k \leqslant p\}$, to the first $p+1$ autocorrelation coefficients $r_h(0), \cdots, r_h(p)$. These $p+1$ equations can be written in matrix form as

$$\begin{bmatrix} r_h(0) & r_h^*(1) & \cdots & r_h^*(p) \\ r_h(1) & r_h(0) & \cdots & r_h^*(p-1) \\ \vdots & \vdots & \ddots & \vdots \\ r_h(p) & r_h(p-1) & \cdots & r_h(0) \end{bmatrix} \begin{bmatrix} 1 \\ a_p(1) \\ \vdots \\ a_p(p) \end{bmatrix} = \begin{bmatrix} |b(0)|^2 \\ 0 \\ \vdots \\ 0 \end{bmatrix} \tag{3.41}$$

If the first $p+1$ autocorrelation values are given, equation (3.41) is a system with $p+1$ linear equations, and the coefficient matrix is a Hermitian, Toeplitz matrix that can be solved for $b(0)$ and $\{a_p(k)\}$.

Because of the special structure in equation (3.41), the model parameters are found from the autocorrelations by using the last set of p equations in equation (3.41), followed by the computation of $b(0)$ from the first equation, which is the same as equation (3.34). From equation (3.41), we can write in matrix notation

$$\boldsymbol{R}_h \mathrm{a}_p = -\boldsymbol{r}_h \tag{3.42}$$

where $\boldsymbol{R}_h$ is the autocorrelation matrix, a_p is the vector of the model parameters, and $\boldsymbol{r}_h$ is the vector of autocorrelations. Since $r_x(l)=\sigma_w^2 r_h(l)$, we can also express the model parameters in terms of the autocorrelation $r_x(l)$ of the output process $x(n)$ as follows:

$$\boldsymbol{R}_x \mathrm{a}_p = -\boldsymbol{r}_x \tag{3.43}$$

These equations are known as the Yule-Walker equations in the statistics literature. In the sequel, we drop the subscript from the autocorrelation sequence or matrix whenever the analysis holds for both the impulse response and the model output.

Because of the Toeplitz structure and the nature of the right-hand side, the linear systems equation (3.42) and equation (3.43) can be solved recursively by using the algorithm of Levinson-Durbin. After a_p is solved, the system gain $b(0)$ can be computed from equation (3.34).

Therefore, given $r(0), r(1), \cdots, r(p)$, we can completely specify the parameters of the all-pole model by solving a set of linear equations. Below, we will see that the converse is also true: Given the model parameters, we can find the first $p+1$ autocorrelations by solving a set of linear equations. This elegant solution of the spectral factorization problem is unique to all-pole models. In the case in which the model contains zeros ($Q\neq 0$), the spectral factorization problem requires the solution of a nonlinear system of equations.

Autocorrelation in terms of model parameters. If we normalize the autocorrelations in equation (3.42) by dividing throughout by $r(0)$, we obtain the following system of equations

$$\boldsymbol{P}\mathrm{a}_p = -\boldsymbol{\rho} \tag{3.44}$$

where $\boldsymbol{P}$ is the normalized autocorrelation matrix and

$$\boldsymbol{\rho} = [\rho(1)\ \rho(2) \cdots \rho(p)]^{\mathrm{H}} \tag{3.45}$$

is the vector of normalized autocorrelations. This set of p equations relates the p model coefficients with the first p (normalized) autocorrelation values. If the poles of the all-pole filter are strictly inside the unit circle, the mapping between the p-dimensional vectors $\boldsymbol{a}_p$ and $\boldsymbol{\rho}$ is unique. If, in fact, we are given the vector $\boldsymbol{a}_p$, then the normalized autocorrelation vector $\boldsymbol{\rho}$ can be computed from $\boldsymbol{a}_p$ by using the set of equations that can be deduced from equation (3.44)

$$\boldsymbol{A}\boldsymbol{\rho} = -\boldsymbol{a}_p \tag{3.46}$$

where $\langle \boldsymbol{A} \rangle_{ij} = a_p(i-j) + a_p(i+j)$, assuming $a_p(m) = 0$ for $m<0$ and $m>p$.

Given the set of coefficients in $\boldsymbol{a}_p$, $\boldsymbol{\rho}$ can be obtained by solving equation (3.46). We will see that, under the assumption of a stable $H(z)$, a solution always exists. If, in addition to $\boldsymbol{a}_p$, we are given $b(0)$, we can evaluate $r(0)$ with equation (3.37) from $\boldsymbol{\rho}$ computed by equation (3.46). Autocorrelation values $r(1)$ for lags $l > p$ are found by using the recursion in equation (3.35) with $r(0), r(1), \cdots, r(p)$.

Correlation matching. All-pole models have the unique distinction that the model parameters are completely specified by the first $p+1$ autocorrelation coefficients via a set of linear equations. We can write

$$\begin{bmatrix} b(0) \\ \boldsymbol{a}_p \end{bmatrix} \leftrightarrow \begin{bmatrix} r(0) \\ \boldsymbol{\rho} \end{bmatrix} \tag{3.47}$$

that is, the mapping of the model parameters $\{b(0), a_p(1), a_p(2), \cdots, a_p(p)\}$ to the autocorrelation coefficients specified by the vector $\{r(0), \rho(1), \rho(2), \cdots, \rho(p)\}$ is reversible and unique. This statement implies that given any set of autocorrelation values $r(0), r(1), \cdots, r(p)$, we can always find an all-pole model whose first $p+1$ autocorrelation coefficients are equal to the given autocorrelations. This correlation matching of all-pole models is quite remarkable. This property is not shared by all-zero models and is true for pole-zero models only under certain conditions, as we will see in Section 3.4.

Spectrum. The z-transform of the autocorrelation $r(1)$ of $H(z)$ is given by

$$R(z) = H(z) H^*\left(\frac{1}{z^*}\right) \tag{3.48}$$

The spectrum is then equal to

$$R(\mathrm{e}^{\mathrm{j}\omega}) = |H(\mathrm{e}^{\mathrm{j}\omega})|^2 = \frac{|b(0)|^2}{|A(\mathrm{e}^{\mathrm{j}\omega})|^2} \tag{3.49}$$

The right-hand side of equation (3.49) suggests a method for computing the spectrum: First compute $A(\mathrm{e}^{\mathrm{j}\omega})$ by taking the Fourier transform of the sequence $\{1, a_p(1), \cdots, a_p(p)\}$, then take the squared of the magnitude and divide $|b(0)|^2$ by the result. The fast Fourier transform (FFT) can be used to this end by appending the sequence $\{1, a_p(1), \cdots, a_p(p)\}$ with as many zeros as needed to compute the desired number of frequency points.

Partial autocorrelation and lattice structures. We have seen that an AP(P) model is completely described by the first $p+1$ values of its autocorrelation. However, we cannot determine the order of the model by using the autocorrelation sequence because it has infinite duration. Suppose that we start fitting models of increasing order m, using the autocorrelation sequence of an AP(P) model and the Yule-Walker equations

$$\begin{bmatrix} 1 & \rho^*(1) & \cdots & \rho^*(m-1) \\ \rho(1) & 1 & \cdots & \vdots \\ \vdots & \vdots & \ddots & \rho^*(1) \\ \rho(m-1) & \cdots & \rho(1) & 1 \end{bmatrix} \begin{bmatrix} a_1^{(m)} \\ a_2^{(m)} \\ \vdots \\ a_m^{(m)} \end{bmatrix} = - \begin{bmatrix} \rho^*(1) \\ \rho^*(2) \\ \vdots \\ \rho^*(m) \end{bmatrix} \tag{3.50}$$

Since $a_m^{(m)} = 0$ for $m > P$, we can use the sequence $a_m^{(m)}, m = 1,2,\cdots$, which is known as the partial autocorrelation sequence (PACS), to determine the order of the all-pole model. We can get

$$a_m^{(m)} = k_m \tag{3.51}$$

that is, the PACS is identical to the lattice parameters.

Furthermore, it has been shown that

$$r(0)\prod_{m=1}^{P}\frac{1-|k_m|}{1+|k_m|} \leqslant R(e^{j\omega}) \leqslant r(0)\prod_{m=1}^{P}\frac{1+|k_m|}{1-|k_m|} \tag{3.52}$$

which indicates that the spectral dynamic range increases if some lattice parameter moves close to 1 or equivalently some pole moves close to the unit circle.

Equivalent model representations. A minimum-phase AP(P) model can be uniquely described by any one of the following representations:

1. direct structure: $\{b(0), a_p(1), a_p(2), \cdots, a_p(p)\}$
2. lattice structure: $\{b(0), k_1, k_2, \cdots, k_p\}$
3. autocorrelation: $\{r(0), r(1), \cdots, r(p)\}$

where we assume, without loss of generality, that $b(0) > 0$. Note that the minimum-phase property requires that all poles be inside the unit circle or all $|k_m| < 1$ or that $\boldsymbol{R}_{p+1}$ be positive definite.

Minimum-phase conditions. If the Toeplitz matrix $\boldsymbol{R}_h$ is positive definite, then $|k_m| < 1$ for all $m = 1,2,\cdots,p$. Therefore, the AP(P) model obtained by solving the Yule-Walker equations is minimum-phase. Therefore, the Yule-Walker equations provide a simple and elegant solution to the spectral factorization problem for all-pole models.

3.2.2 All-Pole Modeling and Linear Prediction

Consider the AP(P) model

$$x(n) = -\sum_{k=1}^{P} a_p(k)x(n-k) + w(n) \tag{3.53}$$

the Mth-order liner predictor of $x(n)$ and the corresponding predictor error $e(n)$ are

$$\hat{x}(n) = -\sum_{k=1}^{M} a_p^0(k)x(n-k) \tag{3.54}$$

$$e(n) = x(n) - \hat{x}(n) = x(n) + \sum_{k=1}^{M} a_p^0(k)x(n-k) \tag{3.55}$$

or

$$x(n) = -\sum_{k=1}^{M} a_p^0(k)x(n-k) + e(n) \tag{3.56}$$

Notice that if the order of the linear predictor equals the order of the all-pole model ($M=P$) and if $a_p^0(k) = a_p(k)$, then the prediction error is equal to the excitation of the all-pole model, that is, $e(n) = w(n)$. Since all-pole modeling and FIR linear prediction are closely related, many properties and algorithms developed for one of them can be applied to the other.

3.2.3 Autoregressive Models

Causal all-pole models excited by white noise play a major role in practical applications and are known as autoregressive (AR) models. An AR(P) model is defined by the difference equation

$$x(n) = -\sum_{k=1}^{p} a_p(k)x(n-k) + w(n) \tag{3.57}$$

where $\{w(n)\} \sim WN(0,\sigma_w^2)$. An AR($P$) model is valid only if the corresponding AP(P) system is stable. In this case, the output $x(n)$ is a stationary sequence with a mean value of zero. Post multiplying equation (3.57) by $x^*(n-l)$ and taking the expectation, we obtain the following recursive relation for the autocorrelation:

$$r_x(l) = -\sum_{k=1}^{p} a_p(k)r_x(l-k) + E\{w(n)x^*(n-l)\} \tag{3.58}$$

Similarly, we can show that $E\{w(n)x^*(n-l)\} = \sigma_w^2 h^*(-l)$. Thus, we have

$$r_x(l) = -\sum_{k=1}^{p} a_p(k)r_x(l-k) + \sigma_w^2 h^*(-l) \text{ for all } l \tag{3.59}$$

The variance of the output signal is

$$\sigma_x^2 = r_x(0) = -\sum_{k=1}^{p} a_p(k)r_x(k) + \sigma_w^2$$

or

$$\sigma_x^2 = \frac{\sigma_w^2}{1+\sum_{k=1}^{p} a_p(k)\rho_x(k)} \tag{3.60}$$

If we substitute $l = 0,1,\cdots,p$ in equation (3.60) and recall that $h(n) = 0$ for $n < 0$, we obtain the following set of Yule-Walker equations:

$$\begin{bmatrix} r_x(0) & r_x(1) & \cdots & r_x(p) \\ r_x^*(1) & r_x(0) & \cdots & r_x(p-1) \\ \vdots & \vdots & \ddots & \vdots \\ r_x^*(p) & r_x^*(p-1) & \cdots & r_x(0) \end{bmatrix} \begin{bmatrix} 1 \\ a_1 \\ \vdots \\ a_p \end{bmatrix} = \begin{bmatrix} \sigma_w^2 \\ 0 \\ \vdots \\ 0 \end{bmatrix} \tag{3.61}$$

Careful inspection of the above equations reveals their similarity to the corresponding relationships developed previously for the AP(P) model. This should be no surprise since the power spectrum of the white noise is flat. However, there is one important difference we should clarify: AP(P) models are specified with a gain $b(0)$ and the parameters $\{a_p(1),a_p(1),\cdots,a_p(p)\}$, but for AR($P$) models we set the gain $b(0)=1$ and define the model by the variance of the white excitation σ_w^2 and the parameters $\{a_p(1),a_p(1),\cdots,a_p(p)\}$. In other words, we incorporate the gain of the model into the power of the input signal. Thus, the power spectrum of the output is $R_x(e^{j\omega}) = \sigma_w^2 |H(e^{j\omega})|^2$. Similar arguments apply to all parametric models driven by white noise. We just rederived some of the relationships to clarify these issues and to provide additional insight into the subject.

3.2.4 Lower-Order Models

In this section, we derive the properties of lower-order all-pole models, namely, first-and second-order models, with real coefficients.

First-order all-pole model: AP (1)

An AP (1) model has a transfer function

$$H(z)=\frac{b(0)}{1+az^{-1}} \tag{3.62}$$

with a single pole at $z=-a$ on the real axis. It is clear that $H(z)$ is minimum-phase if

$$-1<a<1 \tag{3.63}$$

From equation (3.35) with $p=1$ and $l=1$, we have

$$a_p(1)=-\frac{r(1)}{r(0)}=-\rho(1) \tag{3.64}$$

Similarly, from equation (3.51) with $m=1$,

$$a_1^{(1)}=a=-\rho(1)=k_1 \tag{3.65}$$

Since from equation (3.27), $h(0)=b(0)$, and from equation (3.28) $h(n)=-a_p(1)h(n-1)$ for $n>0$, the impulse response of a single-pole filter is given by

$$h(n)=b(0)\ (-a)^n u(n) \tag{3.66}$$

The same result can, of course, be obtained by taking the inverse z-transform of $H(z)$.

The autocorrelation is found in a similar fashion. From equation (3.35) and by using the fact that the autocorrelation is an even function,

$$r(l)=r(0)\ (-a)^{|l|}\ \textit{for all } l \tag{3.67}$$

and from equation (3.37)

$$r(0)=\frac{b\ (0)^2}{1-a^2}=\frac{b\ (0)^2}{1-k_1^2} \tag{3.68}$$

Therefore, if the energy $r(0)$ in the impulse response is set to unity, then the gain must be set to

$$b(0)=\sqrt{1-k_1^2}\qquad r(0)=1 \tag{3.69}$$

The z-transform of the autocorrelation is then

$$R(z)=\frac{b\ (0)^2}{(1+az^{-1})(1+az)}=r(0)\sum_{l=-\infty}^{\infty}(-a)^{|l|}z^{-l} \tag{3.70}$$

and the spectrum is

$$R(\mathrm{e}^{\mathrm{j}\omega})=|H(\mathrm{e}^{\mathrm{j}\omega})|^2=\frac{b\ (0)^2}{|1+a\mathrm{e}^{-\mathrm{j}\omega}|^2}=\frac{b\ (0)^2}{1+2a\cos\omega+a^2} \tag{3.71}$$

Consider now the real-valued AR (1) process $x(n)$ generated by

$$x(n)=-ax(n-1)+w(n) \tag{3.72}$$

where $\{w(n)\}\sim WN(0,\sigma_w^2)$. Using the equation $R_x(z)=\sigma_w^2 H(z)H^*(1/z^*)$ and previous results, we can see that the autocorrelation and the PSD of $x(n)$ are given by

$$R_x(\mathrm{e}^{\mathrm{j}\omega})=\sigma_\omega^2\frac{1-a^2}{1+a^2+2a\cos\omega}$$

and

$$r_x(l)=\frac{\sigma_\theta^2}{1-a^2}(-a)^{|l|}$$

respectively. Since $\sigma_x^2=r_x(0)=\sigma_w^2/(1-a^2)$, the SFM of $x(n)$ is

$$\mathrm{SFM}_x = \frac{\sigma_\omega^2}{\sigma_x^2} = 1 - a^2 \tag{3.73}$$

Clearly, if $a = 0$, then from equation (3.72), $x(n)$ is a white noise process and from equation (3.73), $\mathrm{SFM}_x = 1$. If $a \to 1$, then $\mathrm{SFM}_x \to 0$; and in the limit when $a = 1$, the process becomes a random walk process, which is a nonstationary process with linearly increasing variance $E\{x^2(n)\} = n\sigma_\omega^2$. The correlation matrix is Toeplitz, and it is a rare exception in which eigenvalues and eigenvectors can be described by analytical expressions.

Second-order all-pole model: AP (2)

The system function of an AP (2) model is given by

$$H(z) = \frac{b(0)}{1 + a_1 z^{-1} + a_2 z^{-1}} = \frac{b(0)}{(1 - p_1 z^{-1})(1 - p_2 z^{-1})} \tag{3.74}$$

We have

$$\begin{aligned} a_1 &= -(p_1 + p_2) \\ a_2 &= p_1 p_2 \end{aligned} \tag{3.75}$$

Recall that $H(z)$ is minimum-phase if the two poles p_1 and p_2 are inside the unit circle. Under these conditions, a_1 and a_2 lie in a triangular region defined by

$$\begin{aligned} -1 &< a_2 < 1 \\ a_2 - a_1 &> -1 \\ a_2 + a_1 &> -1 \end{aligned} \tag{3.76}$$

The first condition follows from equation (3.75) since $|p_1| < 1$ and $|p_2| < 1$. The last two conditions can be derived by assuming real roots and setting the larger root to less than 1 and the smaller root to greater than -1. By adding the last two conditions, we obtain the redundant condition $a_2 > -1$.

Complex roots occur in the region

$$\frac{a_1^2}{4} < a_2 \leqslant 1 \quad \textit{complex poles} \tag{3.77}$$

with $a_2 = 1$ resulting in both roots being on the unit circle. Note that, in order to have complex poles, a_2 cannot be negative. If the complex poles are written in polar form

$$p_i = re^{\pm j\theta} \quad 0 \leqslant r \leqslant 1 \tag{3.78}$$

then

$$a_1 = -2r\cos\theta \quad a_2 = r^2 \tag{3.79}$$

and

$$H(z) = \frac{b(0)}{1 - (2r\cos\theta)z^{-1} + r^2 z^{-2}} \quad \textit{complex poles} \tag{3.80}$$

Here, r is the radius (magnitude) of the poles, and θ is the angle or normalized frequency of the poles.

Impulse response. The impulse response of an AP (2) model can be written in terms of its two poles by evaluating the inverse z-transform of equation (3.74). The result is

$$h(n) = \frac{b(0)}{p_1 - p_2}(p_1^{n+1} - p_2^{n+1})u(n) \tag{3.81}$$

for $p_1 \neq p_2$. Otherwise, for $p_1 = p_2 = p$,

$$h(n) = b(0)(n + 1)p^n u(n) \tag{3.82}$$

In the special case of a complex conjugate pair of poles $p_1 = re^{j\theta}$ and $p_2 = re^{-j\theta}$, equation (3.81) reduces to

$$h(n) = b(0)r^n \frac{\sin[(n + 1)\theta]}{\sin \theta}u(n) \quad \textit{complex poles} \tag{3.83}$$

Since $0 < r < 1$, $h(n)$ is a damped sinusoid of frequency θ.

Autocorrelation. The autocorrelation can also be written in terms of the two poles as

$$r(l) = \frac{b(0)^2}{(p_1 - p_2)(1 - p_1 p_2)}\left(\frac{p_1^{l+1}}{1 - p_1^2} - \frac{p_2^{l+1}}{1 - p_2^2}\right) \quad l \geqslant 0 \tag{3.84}$$

from which we can deduce the energy

$$r(0) = \frac{b(0)^2(1 + p_1 p_2)}{(1 - p_1 p_2)(1 - p_1^2)(1 - p_2^2)} \tag{3.85}$$

For the special case of a complex conjugate pole pair, equation (3.84) can be rewritten as

$$r(l) = \frac{b(0)^2 r^l\{\sin[(l + 1)\theta] - r^2 \sin[(l - 1)\theta]\}}{[(1 - r^2)\sin \theta](1 - 2r^2 \cos 2\theta + r^4)} \quad l \geqslant 0 \tag{3.86}$$

Then from equation (3.85) we can write an expression for the energy in terms of the polar coordinates of the complex conjugate pole pair

$$r(0) = \frac{b(0)^2(1 + r^2)}{(1 - r^2)(1 - 2r^2 \cos 2\theta + r^4)} \tag{3.87}$$

The normalized autocorrelation is given by

$$\rho(l) = \frac{r^l\{\sin[l + 1)\theta] - r^2 \sin[(l - 1)\theta]\}}{(1 + r^2)\sin \theta} \quad l \geqslant 0 \tag{3.88}$$

which can be rewritten as

$$\rho(l) = \frac{1}{\cos \beta} r^l \cos(l\theta - \beta) \quad l \geqslant 0 \tag{3.89}$$

where

$$\tan \beta = \frac{(1 - r^2)\cos \theta}{(1 + r^2)\sin \theta} \tag{3.90}$$

Therefore, $\rho(l)$ is a damped cosine wave with its maximum amplitude at the origin.

Spectrum. By setting the two poles equal to

$$p_1 = r_1 e^{j\theta_1} \quad p_2 = r_2 e^{j\theta_2} \tag{3.91}$$

the spectrum of an AP (2) model can be written as

$$R(e^{j\omega}) = \frac{b(0)^2}{[1 - 2r_1\cos(\omega - \theta_1) + r_1^2][1 - 2r_2\cos(\omega - \theta_2) + r_2^2]} \tag{3.92}$$

There are four cases of interest

Pole locations	Peak locations	Type of $R(e^{j\omega})$
$p_1>0, p_2>0$	$\omega=0$	Low-pass
$p_1<0, p_2<0$	$\omega=\pi$	High-pass
$p_1>0, p_2<0$	$\omega=0, \omega=\pi$	Stopband
$p_{1,2}=re^{\pm j\theta}$	$0<\omega<\pi$	Bandpass

and they depend on the location of the poles on the complex plane.

We concentrate on the fourth case of complex conjugate poles, which is of greatest interest. The spectrum is given by

$$R(e^{j\omega})=\frac{b(0)^2}{[1-2r\cos(\omega-\theta)+r^2][1-2r\cos(\omega+\theta)+r^2]} \tag{3.93}$$

The peak of this spectrum can be shown to be located at a frequency w_c, given by

$$\cos\omega_c=\frac{1+r^2}{2r}\cos\theta \tag{3.94}$$

Since $1+r^2>2r$ for $r<1$, and we have

$$\cos\omega_c>\cos\theta \tag{3.95}$$

the spectral peak is lower than the pole frequency for $0<\theta<\pi/2$ and higher than the pole frequency for $\pi/2<\theta<\pi$.

Equivalent model descriptions. We now write explicit formulas for a_1 and a_2 in terms of the lattice parameters k_1 and k_2 and the autocorrelation coefficients. We have

$$\begin{aligned} a_1&=k_1(1+k_2)\\ a_2&=k_2 \end{aligned} \tag{3.96}$$

and the inverse relations

$$\begin{aligned} k_1&=\frac{a_1}{1+a_2}\\ k_2&=a_2 \end{aligned} \tag{3.97}$$

From the Yule-Walker equations equation (3.35), we can write the two equations

$$\begin{aligned} a_p(1)r(0)+a_p(2)r(1)&=-r(1)\\ a_p(1)r(1)+a_p(2)r(0)&=-r(2) \end{aligned} \tag{3.98}$$

which can be solved for $a_p(1)$ and $a_p(2)$ in terms of $\rho(1)$ and $\rho(2)$

$$\begin{aligned} a_1&=-\rho(1)\frac{1-\rho(2)}{1-\rho^2(1)}\\ a_2&=\frac{\rho^2(1)-\rho(2)}{1-\rho^2(1)} \end{aligned} \tag{3.99}$$

or for $\rho(1)$ and $\rho(2)$ in terms of a_1 and a_2

$$\begin{aligned} \rho(1)&=-\frac{a_1}{1+a_2}\\ \rho(2)&=-a_1\rho(1)-a_2=\frac{a_1^2}{1+a_2}-a_2 \end{aligned} \tag{3.100}$$

From the equations above, we can also write the relation and inverse relation between the coefficients k_1 and k_2 and the normalized autocorrelations $\rho(1)$ and $\rho(2)$ as

$$k_1 = -\rho(1)$$
$$k_2 = \frac{\rho^2(1) - \rho(2)}{1 - \rho^2(1)} \tag{3.101}$$

and

$$\rho(1) = -k_1$$
$$\rho(2) = k_1(1 + k_2) - k_2 \tag{3.102}$$

The gain $b(0)$ can also be written in terms of the other coefficients. From equation (3.37), we have

$$b^2(0) = r(0)[1 + a_1\rho(1) + a_2\rho(2)] \tag{3.103}$$

which can be shown to be equal to

$$b^2(0) = r(0)(1 - k_1)(1 - k_2) \tag{3.104}$$

Minimum-phase conditions. In equation (3.76), we have a set of conditions on a_1 and a_2 so that the AP (2) model is minimum-phase. Similar relations and regions can be derived for the other types of parameters, as we will show below. In terms of k_1 and k_2, the AP (2) model is minimum-phase if

$$|k_1| < 1 \quad |k_2| < 1 \tag{3.105}$$

The results in complex roots are specified by

$$0 < k_2 < 1 \tag{3.106}$$

$$k_1^2 < \frac{4k_2}{(1 + k_2)^2} \tag{3.107}$$

Because of the correlation matching property of all-pole models, we can find a minimum-phase all-pole model for every positive definite sequence of autocorrelation values. Therefore, the admissible region of autocorrelation values coincides with the positive definite region. The positive definite condition is equivalent to having all the principal minors of the autocorrelation matrix in equation (3.41) be positive definite; that is, the corresponding determinants are positive. For $p = 2$, there are two conditions:

$$\det\begin{bmatrix} 1 & \rho(1) \\ \rho(1) & 1 \end{bmatrix} > 0 \quad \det\begin{bmatrix} 1 & \rho(1) & \rho(2) \\ \rho(1) & 1 & \rho(1) \\ \rho(2) & \rho(1) & 1 \end{bmatrix} > 0 \tag{3.108}$$

These two conditions reduce to

$$|\rho(1)| < 1 \tag{3.109}$$

$$2\rho^2(1)-1<\rho(2)<1 \tag{3.110}$$

Conditions equation (3.110) can also be derived from equation (3.76) and equation (3.100). The first condition in equation (3.110) is equivalent to

$$\left|\frac{a_1}{1 + a_2}\right| < 1 \tag{3.111}$$

which can be shown to be equivalent to the last two conditions in equation (3.76).

3.3 All-Zero Models

In this section, we investigate the properties of the all-zero model. The output of the all-zero model is the weighted average of delayed versions of the input signal

$$x(n) = \sum_{k=0}^{q} d_k w(n-k) \tag{3.112}$$

where q is the order of the model. The system function is

$$H(z) = D(z) = \sum_{k=0}^{q} d_k z^{-k} \tag{3.113}$$

The all-zero model can be implemented by using either a direct or a lattice structure. Notice that the same set of parameters can be used to implement either an all-zero or an all-pole model by using a different structure.

3.3.1 Model Properties

We next provide a brief discussion of the properties of the all-zero model.

Impulse response. It can be easily seen that the AZ (Q) model is an FIR system with an impulse response

$$h(n) = \begin{cases} d_n & 0 \leqslant n \leqslant q \\ 0 & \text{elsewhere} \end{cases} \tag{3.114}$$

Autocorrelation. The autocorrelation of the impulse response is given by

$$r_h(l) = \sum_{n=-\infty}^{\infty} h(n)h^*(n-l) = \begin{cases} \sum_{k=0}^{q-1} d_k d_{k+l}^* & 0 \leqslant l \leqslant q \\ 0 & l > q \end{cases} \tag{3.115}$$

and

$$r_h^*(-l) = r_h(l) \ \ all\ l \tag{3.116}$$

We usually set $d_0 = 1$, which implies that

$$r_h(l) = d_l^* + d_1 d_{l+1}^* + \cdots + d_{q-1} d_q^* \qquad l = 0,1,\cdots,q \tag{3.117}$$

hence, the normalized autocorrelation is

$$\rho_h(l) = \begin{cases} \dfrac{d_l^* + d_1 d_{l+1}^* + \cdots + d_{q-1} d_q^*}{1 + |d_1|^2 + \cdots + |d_q|^2} & l = 1,2,\cdots,q \\ 0 & l > q \end{cases} \tag{3.118}$$

We see that the autocorrelation of an AZ(Q) model is zero for lags $|l|$ exceeding the order q of the model. If $\rho_h(1), \rho_h(2), \cdots, \rho_h(q)$ are known, then the q equation (3.118) can be solved for model parameters $d_1, d_2, \cdots, d_q$. However, unlike the Yule-Walker equations for the AP(P) model, which are linear, equation (3.118) is nonlinear and their solution is quite complicated.

Spectrum. The spectrum of the AZ(Q) model is given by

$$R_h(e^{j\omega}) = D(z)D(z^{-1})|_{z=e^{j\omega}} = |D(e^{j\omega})|^2 = \sum_{l=-q}^{q} r_h(l)e^{-j\omega l} \tag{3.119}$$

which is basically a trigonometric polynomial.

Impulse train excitations. The response $\tilde{h}(n)$ of the AZ(Q) model to a periodic impulse train with period L is periodic with the same period, and its spectrum is a sampled version of equation (3.119) at multiples of $2\pi/L$. Therefore, to recover the autocorrelation $r_h(l)$ and the spectrum $R_h(e^{j\omega})$ from the autocorrelation or spectrum of $\tilde{h}(n)$, we should have $L \geqslant 2q+1$ in order to avoid aliasing in the autocorrelation lag domain. Also, if $L \geqslant q$, the impulse response $h(n)$, $0 \leqslant n \leqslant q$, can be recovered from the response $\tilde{h}(n)$ (no time-domain aliasing).

Partial autocorrelation and lattice-ladder structures. The PACS of an AZ(Q) model is computed by fitting a series of AP(P) models for $p = 1, 2, \cdots$, to the autocorrelation sequence equation (3.118) of the AZ(Q) model. Since the AZ(Q) model is equivalent to an AP (∞) model, the PACS of an all-zero model has infinite extent and behaves as the autocorrelation sequence of an all-pole model. This is illustrated later for the low-order AZ (1) and AZ (2) models.

3.3.2 Moving-Average Models

A moving-average model is an AZ(Q) model with $d_0 = 1$ driven by white noise, that is,

$$x(n) = w(n) + \sum_{k=1}^{q} d_k w(n-k) \tag{3.120}$$

where $\{w(n)\} \sim WN(0, \sigma_w^2)$. The output $x(n)$ has zero mean and variance of

$$\sigma_x^2 = \sigma_w^2 \sum_{k=0}^{q} |d_k|^2 \tag{3.121}$$

The autocorrelation and power spectrum are given by $r_x(l) = \sigma_w^2 r_h(l)$ and $R_x(e^{j\omega}) = \sigma_\omega^2 |D(e^{j\omega})|^2$, respectively. Clearly, observations that are more than q samples apart are uncorrelated because the autocorrelation is zero after lag q.

3.3.3 Lower-Order Models

To familiarize ourselves with all-zero models, we next investigate in detail the properties of the AZ (1) and AZ (2) models with real coefficients.

The first-order all-zero model: AZ (1). For generality, we consider an AZ (1) model whose system function is

$$H(z) = G(1 + d_1 z^{-1}) \tag{3.122}$$

The model is stable for any value of d_1 and minimum-phase for $-1 < d_1 < 1$. The autocorrelation is the inverse z-transform of

$$R_h(z) = H(z)H(z^{-1}) = G^2[d_1 z + (1 + d_1^2) + d_1 z^{-1}] \tag{3.123}$$

Hence, $r_h(0) = G^2(1 + d_1^2)$, $r_h(1) = r_h(-1) = G^2 d_1$, and $r_h(l) = 0$ elsewhere. Therefore, the normalized autocorrelation is

$$\rho_h(l) = \begin{cases} 1 & l = 0 \\ \dfrac{d_1}{1 + d_1^2} & l = \pm 1 \\ 0 & |l| \geq 2 \end{cases} \tag{3.124}$$

The condition $-1 < d_1 < 1$ implies that $|\rho_h(1)| \leqslant \frac{1}{2}$ for a minimum-phase model. From $\rho_h(1) = d_1/(1 + d_1^2)$, we obtain the quadratic equation

$$\rho_h(1) d_1^2 - d_1 + \rho_h(1) = 0 \tag{3.125}$$

which has the following two roots:

$$d_1 = \frac{1 \pm \sqrt{1 - 4\rho_h^2(1)}}{2\rho_h(1)} \tag{3.126}$$

Since the product of the roots is 1, if d_1 is a root, then $1/d_1$ must also be a root. Hence, only one of these two roots can satisfy the minimum-phase condition $-1 < d_1 < 1$.

The spectrum is obtained by setting $z = e^{j\omega}$ in equation (3.123), or from equation (3.119)

$$R_h(e^{j\omega}) = G^2(1 + d_1^2 + 2d_1 \cos \omega) \tag{3.127}$$

The autocorrelation is positive definite if $R_h(e^{j\omega}) > 0$, which holds for all values of d_1. Note that if $d_1 > 0$, then $\rho_h(1) > 0$ and the spectrum has low-pass behavior, whereas a high-pass spectrum is obtained when $d_1 < 0$.

The first lattice parameter of the AZ (1) model is $k_1 = -\rho(1)$. The PACS can be obtained from the Yule-Walker equations by using the autocorrelation sequence equation (3.124). Indeed, after some algebra we obtain

$$k_m = \frac{(-d_1)^m (1 - d_1^2)}{1 - d_1^{2(m+1)}} \qquad m = 1, 2, \cdots, \infty \tag{3.128}$$

Notice the duality between the ACS and PACS of AP (1) and AZ (1) models.

Consider now the MA (1) real-valued process $x(n)$ generated by

$$x(n) = w(n) + bw(n-1)$$

where $\{w(n)\} \sim WN(0, \sigma_w^2)$. Using $R_x(z) = \sigma_w^2 H(z) H(z^{-1})$, we obtain the PSD function

$$R_x(e^{j\omega}) = \sigma_w^2(1 + b^2 + 2b\cos \omega)$$

which has low-pass (high-pass) characteristics if $0 < b \leqslant 1$ $(-1 \leqslant b < 0)$. Since $\sigma_x^2 = r_x(0) = \sigma_w^2(1 + b^2)$, we have

$$\text{SFM}_x = \frac{\sigma_w^2}{\sigma_x^2} = \frac{1}{1 + b^2} \tag{3.129}$$

which is maximum for $b = 0$ (white noise). The correlation matrix is banded Toeplitz (only a number of diagonals close to the main diagonal are nonzero)

$$\boldsymbol{R}_x = \sigma_w^2(1 + b^2) \begin{bmatrix} 1 & b & 0 & \cdots & 0 \\ b & 1 & b & \cdots & 0 \\ 0 & b & 1 & \cdots & 0 \\ \vdots & \vdots & \vdots & \ddots & \vdots \\ 0 & 0 & 0 & \cdots & 1 \end{bmatrix} \tag{3.130}$$

and its eigenvalues and eigenvectors are given by $\lambda_k = R_x(e^{j\omega_k})$, $q_n^{(k)} = \sin \omega_k n$, $\omega_k = \pi k/(M+1)$, where $k = 1, 2, \cdots, M$.

The second-order all-zero model: AZ (2). Now let us consider the second-order all-zero model. The system function of the AZ (2) model is

$$H(z) = G(1 + d_1 z^{-1} + d_2 z^{-2}) \tag{3.131}$$

The system is stable for all values of d_1 and d_2, and minimum-phase (see the discussion for the AP (2) model) if

$$\begin{aligned} -1 < d_2 < 1 \\ d_2 - d_1 > -1 \\ d_2 + d_1 > -1 \end{aligned} \tag{3.132}$$

The normalized autocorrelation and the spectrum are

$$\rho_h(l) = \begin{cases} 1 & l = 0 \\ \dfrac{d_1(1 + d_2)}{1 + d_1^2 + d_2^2} & l = \pm 1 \\ \dfrac{d_2}{1 + d_1^2 + d_2^2} & l = \pm 2 \\ 0 & |l| \geqslant 3 \end{cases} \tag{3.133}$$

and

$$R_h(e^{j\omega}) = G[(1 + d_1^2 + d_2^2) + 2d_1(1 + d_2)\cos \omega + 2d_2 \cos 2\omega] \tag{3.134}$$

respectively.

The minimum-phase region in the autocorrelation domain is described by the equations

$$\begin{aligned} \rho(2) + \rho(1) &= -0.5 \\ \rho(2) - \rho(1) &= -0.5 \\ \rho^2(1) &= 4\rho(2)[1 - 2\rho(2)] \end{aligned} \tag{3.135}$$

The formula for the PACS is quite involved. The important thing is the duality between the ACS and the PACS of AZ (2) and AP (2) models.

3.4 Pole-Zero Models

We will focus on causal pole-zero models with a recursive input-output relationship given by

$$x(n) = -\sum_{k=1}^{p} a_k x(n-k) + \sum_{k=0}^{q} d_k w(n-k) \tag{3.136}$$

where we assume that $p > 0$ and $q \geqslant 1$. The models can be implemented using either direct-form or lattice-ladder structures.

3.4.1 Model Properties

In this section, we present some of the basic properties of pole-zero models.

Impulse response. The impulse response of a causal pole-zero model can be written in recursive form equation (3.136) as

$$h(n) = -\sum_{k=1}^{p} a_k h(n-k) + d_n \qquad n \geqslant 0 \tag{3.137}$$

where

$$d_n = 0 \qquad n > q$$

and $h(n) = 0$ for $n < 0$. Clearly, this formula is useful if the model is stable. From equation (3.137), it is clear that

$$h(n) = -\sum_{k=1}^{p} a_k h(n-k) \qquad n > q \tag{3.138}$$

so that the impulse response obeys the linear prediction equation for $n > q$. Thus if we are given $h(n)$, $0 \leqslant n \leqslant p+q$, we can compute $\{a_k\}$ from equation (3.138) by using the p equations specified by $q+1 \leqslant n \leqslant q+p$. Then we can compute $\{d_k\}$ from equation (3.137), using $0 \leqslant n \leqslant q$. Therefore, the first $p+q+1$ values of the impulse response completely specify the pole-zero model.

If the model is minimum-phase, the impulse response of the inverse model $h_I(n) = Z^{-1}\{A(z)/D(z)\}$, $d_0 = 1$ can be computed in a similar manner.

Autocorrelation. The complex spectrum of $H(z)$ is given by

$$R_h(z) = H(z)H^*\left(\frac{1}{z^*}\right) = \frac{D(z)D^*(1/z^*)}{A(z)A^*(1/z^*)} \triangleq \frac{R_d(z)}{R_a(z)} \tag{3.139}$$

where $R_d(z)$ and $R_a(z)$ are both finite two-sided polynomials. In a manner similar to the all-pole case, we can write a recursive relation between the autocorrelation, impulse response, and parameters of the model. Indeed, from equation (3.139) we obtain

$$A(z)R_h(z) = D(z)H^*\left(\frac{1}{z^*}\right) \tag{3.140}$$

Taking the inverse z-transform of equation (3.140) and noting that the inverse z-transform of $H^*(1/z^*)$ is $h^*(-n)$, we have

$$\sum_{k=0}^{p} a_k r_h(l-k) = \sum_{k=0}^{q} d_k h^*(k-l) \text{ for all } l \tag{3.141}$$

Since $h(n)$ is causal, we see that the right-hand side of equation (3.141) is zero for $l > q$:

$$\sum_{k=0}^{p} a_k r_h(l-k) = 0 \qquad l > q \tag{3.142}$$

Therefore, the autocorrelation of a pole-zero model obeys the linear prediction equation for $l > q$.

Because the impulse response $h(n)$ is a function of a_k and d_k, the set of equations in equation (3.141) is nonlinear in terms of parameters a_k and d_k. However, equation (3.142) is linear in a_k; therefore, we can compute $\{a_k\}$ from equation (3.142), using the set of equations for $l = q+1, \cdots, q+p$, which can be written in matrix form as

$$\begin{bmatrix} r_h(q) & r_h(q-1) & \cdots & r_h(q-p+1) \\ r_h(q+1) & r_h(q) & \cdots & r_h(q-p+2) \\ \vdots & \vdots & \ddots & \vdots \\ r_h(q+p-1) & r_h(q+p-2) & \cdots & r_h(q) \end{bmatrix} \begin{bmatrix} a_1 \\ a_2 \\ \vdots \\ a_p \end{bmatrix} = -\begin{bmatrix} r_h(q+1) \\ r_h(q+2) \\ \vdots \\ r_h(q+p) \end{bmatrix} \tag{3.143}$$

or

$$\bar{R}_h a = -\bar{r}_h \tag{3.144}$$

Here, $\bar{R}_h$ is a non-Hermitian Toeplitz matrix, and the linear system equation (3.143) can be solved by using the algorithm of Trench's (Trench in 1964; Carayannis in 1981).

Even after we solve for a, equation (3.141) continues to be nonlinear in d_k. To compute d_k, we use equation (3.139) to find $R_d(z)$

$$R_d(z) = R_a(z) R_h(z) \tag{3.145}$$

where the coefficients of $R_a(z)$ are given by

$$r_a(l) = \sum_{k=k_1}^{k=k_2} a_k a_{k+|l|}^*, \ -p \leqslant l \leqslant p, k_1 = \begin{cases} 0, l \geqslant 0 \\ -l, l < 0 \end{cases}, k_2 = \begin{cases} p-l, & l \geqslant 0 \\ p, & l < 0 \end{cases} \tag{3.146}$$

From equation (3.145), $r_d(l)$ is the convolution of $r_a(l)$ with $r_h(l)$, given by

$$r_d(l) = \sum_{k=-p}^{p} r_a(k) r_h(l-k) \tag{3.147}$$

If $r(l)$ was originally the autocorrelation of a PZ (P, Q) model, then $r_a(l)$ in equation (3.147) will be zero for $|l| > q$. Since $R_d(z)$ is specified, it can be factored into the product of two polynomials $D(z)$ and $D^*(1/z^*)$, where $D(z)$ is minimum-phase.

Therefore, we have seen that, given the values of the autocorrelation $r_h(l)$ of a PZ(P,Q) model in the range $0 \leqslant l \leqslant p+q$, we can compute the values of the parameters $\{a_k\}$ and $\{d_k\}$ such that $H(z)$ is minimum-phase. Now, given the parameters of a pole-zero model, we can compute its autocorrelation as follows. Equation (3.139) can be written as

$$R_h(z) = R_a^{-1}(z) R_d(z) \tag{3.148}$$

where $R_a^{-1}(z)$ is the spectrum of the all-pole model $1/A(z)$, that is, $1/R_a(z)$. The coefficients of $R_a^{-1}(z)$ can be computed from $\{a_k\}$ by using equation (3.37) and equation (3.35). The coefficients of $R_d(z)$ are computed from equation (3.119). Then $R_h(z)$ is the convolution of the two autocorrelations thus computed, which is equivalent to multiplying the two polynomials in equation (3.148) and equating equal powers of z on both sides of the equation. Since $R_d(z)$ is finite, the summations used to obtain the coefficients of $R_h(z)$ are also finite.

Spectrum. The spectrum of $H(z)$ is given by

$$R_h(e^{j\omega}) = |H(e^{j\omega})|^2 = \frac{|D(e^{j\omega})|^2}{|A(e^{j\omega})|^2} \tag{3.149}$$

Therefore, $R_h(e^{j\omega})$ can be obtained by dividing the spectrum of $D(z)$ by the spectrum of $A(z)$. Again, the FFT can be used to advantage in computing the numerator and denominator of equation (3.148). If the spectrum $R_h(e^{j\omega})$ of a PZ (P, Q) model is given, then the parameters of the (minimum-phase) model can be recovered by first computing the autocorrelation $r_h(l)$ as the inverse Fourier transform of $R_h(e^{j\omega})$ and then using the procedure outlined in the previous section to compute the sets of coefficients $\{a_k\}$ and $\{d_k\}$.

Partial autocorrelation and lattice-ladder structures. Since a PZ(P, Q) model is equivalent to an $AP(\infty)$ model, its PACS has infinite extent and behaves, after a certain lag, as the PACS of an all-zero model.

3.4.2 Autoregressive Moving-Average Models

The autoregressive moving-average model is a PZ (*P*, Q) model driven by white noise and is denoted by ARMA (*P*, Q). Again, we set $d_0 = 1$ and incorporate the gain into the variance (power) of the white noise excitation. Hence, a causal ARMA (*P*, Q) model is defined by

$$x(n) = -\sum_{k=1}^{p} a_k x(n-k) + w(n) + \sum_{k=1}^{q} d_k w(n-k) \tag{3.150}$$

where $\{w(n)\} \sim WN(0,\sigma_w^2)$. The ARMA (*P*,Q) model parameters are $\{\sigma_w^2, a_1, \cdots, a_p, d_1, \cdots, d_q\}$. The output has zero mean and variance of

$$\sigma_x^2 = -\sum_{k=1}^{p} a_k r_x(k) + \sigma_w^2 [1 + \sum_{k=1}^{q} d_k h(k)] \tag{3.151}$$

where $h(n)$ is the impulse response of the model. The presence of $h(n)$ in equation (3.151) makes the dependence of σ_x^2 on the model parameters highly nonlinear. The autocorrelation of $x(n)$ is given by

$$\sum_{k=0}^{p} a_k r_x(l-k) = \sigma_w^2 [1 + \sum_{k=1}^{q} d_k h(k-l)] \text{ for all } l \tag{3.152}$$

and the power spectrum by

$$R_x(e^{j\omega}) = \sigma_w^2 \frac{|D(e^{j\omega})|^2}{|A(e^{j\omega})|^2} \tag{3.153}$$

The significance of ARMA (*P*, Q) models is that they can provide more accurate representations than AR or MA models with the same number of parameters. The ARMA model is able to combine the spectral peak matching of the AR model with the ability of the MA model to place nulls in the spectrum.

3.4.3 The First-Order Pole-Zero Model: PZ(1, 1)

Consider the PZ(1, 1) model with the following system function

$$H(z) = G \frac{1 + d_1 z^{-1}}{1 + a_1 z^{-1}} \tag{3.154}$$

where d_1 and a_1 are real coefficients. The model is minimum-phase if

$$\begin{aligned} -1 < d_1 < 1 \\ -1 < a_1 < 1 \end{aligned} \tag{3.155}$$

For the minimum-phase case, the impulse responses of the direct and the inverse models are

$$h(n) = Z^{-1}\{H(z)\} = \begin{cases} 0 & n < 0 \\ G & n = 0 \\ G(-a_1)^{n-1}(d_1 - a_1) & n > 0 \end{cases} \tag{3.156}$$

and

$$h_I(n) = Z^{-1}\{\frac{1}{H(z)}\} = \begin{cases} 0 & n < 0 \\ G & n = 0 \\ G(-d_1)^{n-1}(a_1 - d_1) & n > 0 \end{cases} \tag{3.157}$$

respectively. We note that as the pole $p = -a_1$ gets closer to the unit circle, the impulse response decays more slowly and the model has "longer memory." The zero $z = -d_1$ controls the impulse response of the inverse model in a similar way. The PZ (1, 1) model is equivalent to the AZ (∞) model

$$x(n) = Gw(n) + G\sum_{k=1}^{\infty} h(k)w(n-k) \tag{3.158}$$

or the AP (∞) model

$$x(n) = -\sum_{k=1}^{\infty} h_I(k)x(n-k) + Gw(n) \tag{3.159}$$

If we wish to approximate the PZ (1, 1) model with a finite-order AZ(Q) model, the order q required to achieve a certain accuracy increases as the pole moves closer to the unit circle. Likewise, in the case of an AP(P) approximation, better fits to the PZ (P, Q) model require an increased order p as the zero moves closer to the unit circle.

To determine the autocorrelation, we recall from equation (3.141) that for a causal model

$$r_h(l) = -a_1 r_h(l-1) + Gh(-l) + Gd_1 h(1-l) \quad all\ l \tag{3.160}$$

or

$$\begin{aligned} r_h(0) &= -a_1 r_h(1) + G + Gd_1(d_1 - a_1) \\ r_h(1) &= -a_1 r_h(0) + Gd_1 \\ r_h(l) &= -a_1 r_h(l-1) \quad l \geqslant 2 \end{aligned} \tag{3.161}$$

Solving the first two equations for $r_h(0)$ and $r_h(1)$, we obtain

$$r_h(0) = G\frac{1 + d_1^2 - 2a_1 d_1}{1 - a_1^2} \tag{3.162}$$

$$r_h(1) = G\frac{(d_1 - a_1)(1 - a_1 d_1)}{1 - a_1^2} \tag{3.163}$$

The normalized autocorrelation is given by

$$\rho_h(1) = \frac{(d_1 - a_1)(1 - a_1 d_1)}{1 + d_1^2 - 2a_1 d_1} \tag{3.164}$$

and

$$\rho_h(l) = (-a_1)^{l-1}\rho_h(l-1) \qquad l \geqslant 2 \tag{3.165}$$

Note that given $\rho_h(1)$ and $\rho_h(2)$, we have a nonlinear system of equations that must be solved to obtain a_1 and d_1. By using equation (3.155), equation (3.164), and equation (3.165), it can be shown that the PZ (1,1) is minimum-phase if the ACS satisfies the conditions

$$\begin{aligned} &|\rho(2)| < |\rho(1)| \\ &\rho(2) > \rho(2)[2\rho(1) + 1] \qquad \rho(1) < 0 \\ &\rho(2) > \rho(1)[2\rho(1) - 1] \qquad \rho(1) > 0 \end{aligned} \tag{3.166}$$

3.4.4 Summary and Dualities

Table 3.1 summarizes the key properties of all-zero, all-pole, and pole-zero models. These properties help to identify models for empirical discrete-time signals. Furthermore, the table shows the duality between AZ and AP models. More specifically, we see that

1. An invertible AZ(q) model is equivalent to an AP (∞) model. Thus, it has a finite-extent autocorrelation and an infinite-extent partial autocorrelation.

2. A stable AP (p) model is equivalent to an AZ (∞) model. Thus, it has an infinite-extent autocorrelation and a finite-extent partial autocorrelation.

3. The autocorrelation of an AZ(q) model behaves as the partial autocorrelation of an AP (p) model, and vice versa.

4. The spectra of an AP (p) model and an AZ (q) model are related through an inverse relationship.

These dualities and properties have been shown and illustrated for low-order models in the previous sections.

Table 3.1 Summary of all-pole, all-zero, and pole-zero model properties

Model	AP(P)	AZ(Q)	PZ (P, Q)
Input-output description	$x(n) + \sum_{k=1}^{p} a_k x(n-k) = w(n)$	$x(n) = d_0 w(n) + \sum_{k=1}^{q} d_k w(n-k)$	$x(n) + \sum_{k=1}^{p} a_k x(n-k) = d_0 w(n) + \sum_{k=1}^{q} d_k w(n-k)$
System function	$H(z) = \frac{1}{A(z)} = \frac{d_0}{1 + \sum_{k=1}^{p} a_k z^{-k}}$	$H(z) = D(z) = d_0 + \sum_{k=1}^{q} d_k z^{-k}$	$H(z) = \frac{D(z)}{A(z)}$
Recursive representation	Finite summation	Infinite summation	Infinite summation
Nonrecursive representation	Infinite summation	Finite summation	Infinite summation
Stablity conditions	Poles inside unit circle	Always	Poles inside unit circle
Invertiblity conditions	Always	Zeros inside unit circle	Zeros inside unit circle
Autocorrelation sequence	Infinite duration (damped exponentials and/or sine waves)	Finite summation	Infinite duration (damped exponentials and/or sine waves after $q-p$ lags)
Partial autocorrelation	Tails off Finite duration	Cuts off Infinite duration (damped exponentials and/or sine waves)	Tails off Infinite duration (dominated by damped exponentials and/or sine waves after $q-p$ lags)
Spectrum	Cuts off Good peak matching	Tails off Good "notch" matching	Good peak and valley matching

3.5 Summary

In this chapter, we introduced the classes of pole-zero signal models and discuss their properties. Each model consists of two components: an excitation source and a system. We emphasized that the properties of a signal model depend on the properties of both components. For uncorrelated random inputs, which by definition are excitations of the ARMA model, the second-order moments of the signal model and its minimum phase properties are entirely determined by the system.

We mainly discussed how to determine AP, AZ and PZ models. The parameters of the AP models are determined by the first $p+1$ values of the impulse response. Its autocorrelation sequence can be represented by the inverse z transform of $R_h(z)$. The advantage of an all-pole model is that the normal equations are Toeplitz and may therefore be easily solved using the Levinson-Durbin recursion. The Toeplitz structure also provides an advantage in terms of storage and reduces the amount of computations required to evaluate the autocorrelations in the normal equations. The pole-zero model is determined by the corresponding first $p+q+1$ values of the impulse response. The nomal equations of the pole-zero model are non-Hermitian Toeplitz matrix, and the parameters are solved by the Trench algorithm.

Besides, we described in detail the autocorrelation, power spectral density, and partial correlation of all AZ, AP and PZ models in the general case and first- and second-order models. An understanding of these properties is important for model selection in practical applications.

Exercises

1. The known autoregressive signal model AR(3) is

$$x(n) = \frac{14}{24}x(n-1) + \frac{9}{24}x(n-2) - \frac{1}{24}x(n-3) + w(n)$$

where $\{w(n)\} \sim WN(0,\sigma_w^2)$, $\sigma_w^2 = 1$.

(1) find autocorrelation sequences $R_{xx}(m)$, $m = 0,1,2,3,4,5$

(2) use the autocorrelation sequences from (a) to estimate the parameters $\{\hat{a}_k\}$ of AP(3) and the variance of the input white noise $\hat{\sigma}_w^2$

Solution

for the AP(3) model with real coefficients we have

$$\begin{bmatrix} r_x(0) & r_x(1) & \cdots & r_x(p) \\ r_x(1) & r_x(0) & \cdots & r_x(p-1) \\ \vdots & \vdots & \ddots & \vdots \\ r_x(p) & r_x(p-1) & \cdots & r_x(0) \end{bmatrix} \begin{bmatrix} 1 \\ a_p(1) \\ \vdots \\ a_p(p) \end{bmatrix} = \begin{bmatrix} {\sigma_w}^2 \\ 0 \\ \vdots \\ 0 \end{bmatrix}$$

We rewrite the equation above as $R_x a = r$

Cause autocorrelation function is symmetry. Rewrite the equation above, we have

$$\begin{bmatrix} 1 & a_p(1) & a_p(2) & \cdots & a_p(p-1) & a_p(p) \\ a_p(1) & 1 + a_p(2) & a_p(3) & \cdots & a_p(p) & 0 \\ \vdots & \vdots & \vdots & \ddots & \vdots & \vdots \\ a_p(p) & a_p(p-1) & a_p(p-2) & \cdots & a_p(1) & 1 \end{bmatrix} \begin{bmatrix} r_x(0) \\ r_x(1) \\ \vdots \\ r_x(p) \end{bmatrix} = \begin{bmatrix} {\sigma_w}^2 \\ 0 \\ \vdots \\ 0 \end{bmatrix}$$

We rewrite the equation above as $Ar_x = r$

In this example, $a_p(1) = -\frac{14}{24}, a_p(2) = \frac{9}{24}, a_p(1) = \frac{1}{24}$

The code as follows:

```
clc;
a=[-14/24-9/24 1/24];
b=[1;zeros(3,1)];
A=[1,a(1),a(2),a(3);
a(1),1+a(2),a(3),0;
a(2),a(1)+a(3),1,0;
a(3),a(2),a(1),1];
Rx_T=A \ b
R_T=[Rx_T(1),Rx_T(2),Rx_T(3),Rx_T(4);
Rx_T(2),Rx_T(1),Rx_T(2),Rx_T(3);
Rx_T(3),Rx_T(2),Rx_T(1),Rx_T(2);
Rx_T(4),Rx_T(3),Rx_T(2),Rx_T(1)];
x_T=fsolve(@(x_T)R_T*[1 x_T(1) x_T(2) x_T(3)].'-[x_T(4) 0 0 0].',rand(1,4))
```

The result of the operation is as follows:

Where $\{\hat{a}_k\} = [-0.583,3-0.375,0\ 0.041,7]$ and $\hat{\sigma}_w^2 = 1.000$

2. A PZ(2,2) model is given by $\boldsymbol{x} = [2,1,0,-1,0,1,-1,0,1\cdots]^{\mathrm{T}}$ i. e. $x(0)=2, x(1)=1, x(2)=0$ and so on. In other words, using an approximation of the form

$$H(z) = \frac{b(0) + b(1)z^{-1} + b(2)z^{-2}}{1 + a(1)z^{-1} + a(2)z^{-2}}$$

find the coefficients $b(0), b(1), b(2), a(1)$ and $a(2)$

Solution

From pade equations, we have

$$\begin{bmatrix} 2 & 0 & 0 \\ 1 & 2 & 0 \\ 0 & 1 & 2 \\ -1 & 0 & 1 \\ 0 & -1 & 0 \end{bmatrix} \begin{bmatrix} 1 \\ a(1) \\ a(2) \end{bmatrix} = \begin{bmatrix} b(0) \\ b(1) \\ b(2) \\ 0 \\ 0 \end{bmatrix}$$

Then we use the last two equations

$$\begin{bmatrix} -1 & 0 & 1 \\ 0 & -1 & 0 \end{bmatrix} \begin{bmatrix} 1 \\ a(1) \\ a(2) \end{bmatrix} = \begin{bmatrix} 0 \\ 0 \end{bmatrix}$$

to solve $a(1)$ and $a(1)$.

Using the first three equations, we may solve other coefficients.

Therefore, the model is $H(z) = \dfrac{2 + z^{-2} + 2z^{-2}}{1 + z^{-2}}$

```
Rx_T=
    4.937,7
    4.328,7
    4.196,4
    3.865,4
x_T=
    -0.583,3   -0.375,0   0.041,7   1.000,0
```

References

[1] Brockwell, P. J. , and R. A. Davis. Time Series: Theory and Methods. Springer-Verlag, 1987.

[2] Cryer, Jonathan D. , and Kung-Sik Chan. Time Series Analysis with Applications in r. 2nd ed. Springer, 2008.

[3] J. R. Deller, J. G. Proakis, and J. H. L. Hansen, Discrete-time Processing of Speech Signals. MacMillan, 1993.

[4] J. Durbin. Efficient estimation of parameters in moving-average models. Biometrica, vol. 46, pp. 306–316, 1959.

[5] R. A. Roberts and C. T. Mullis, Digital Signal Processing, Addison Wesley, Reading, MA. 1987.

[6] Trench, William F. "An algorithm for the inversion of finite Toeplitz matrices." Journal of the Society for Industrial and Applied Mathematics 12.3 (1964): 515–522.

Chapter 4

Spectrum Estimation

4.1 Introduction

We consider the estimation of the power spectral density of a wide-sense stationary random process in this chapter. As discussed before, the power spectrum is the Fourier transform of the autocorrelation sequence. Thus, estimating the power spectrum is equivalent to estimating the autocorrelation.

If $x(n)$ is known for all n, estimating the power spectrum is straightforward, in theory, since all that must be done is to determine the autocorrelation sequence $r_x(k) = \lim_{N\to\infty}\{(1/2N+1)\sum_{n=-N}^{N} x(n+k)x^*(n)\}$, and then compute its Fourier transform. But the amount of data that one needs to be processed is never unlimited, and in many cases it may be very small. The second difficulty is that the data is often corrupted by noise or contaminated with an interfering signal. Thus, spectrum estimation is a problem that involves estimating $P_x(e^{j\omega})$ from a finite number of noisy measurements of $x(n)$.

Spectrum estimation is an issue that is important in a variety of different areas and applications. It can be known from the following Chapter 5, for example, that the frequency response of a noncausal Wiener smoothing filter is

$$H(e^{j\omega}) = \frac{P_d(e^{j\omega})}{P_d(e^{j\omega}) + P_v(e^{j\omega})}$$

where $P_d(e^{j\omega})$ is the power spectrum of $d(n)$, the desired output of the Wiener filter, and $P_v(e^{j\omega})$ is the power spectrum of the noise $v(n)$. Therefore, before a Wiener smoothing filter can be designed and implemented, the power spectrum of both $d(n)$ and $v(n)$ must be determined. Since these power spectral densities are not generally known a priori, one is faced with the problem of estimating them from measurements. Another application in which spectrum estimation plays an important role is signal detection and tracking. For example, suppose that a sonar array is placed on the ocean floor to listen for the narrowband acoustic signals that are generated by the rotating machinery or propellers of a ship. Once a narrowband signal is detected, the problem of interest is to estimate its center frequency in order to determine the ships direction or velocity. Since these narrowband signals are typically recorded in a very noisy environment, signal detection and frequency estimation are nontrivial problems that require robust, high-resolution spectrum estimation techniques. Other applications of spectrum estimation include harmonic analysis and prediction, time series extrapolation and interpolation, spectral smoothing, bandwidth compression and beam-forming.

The approaches for spectrum estimation may be generally categorized into two classes. The first includes the classical or *nonparametric* methods that begin by estimating the autocorrelation sequence $r_x(k)$ from a given set of data. The power spectrum is then estimated by Fourier transforming the estimated autocorrelation sequence. The second class includes the nonclassical or parametric approaches, which are based on using a model for the process in order to estimate the power spectrum. One of the limitations of the nonparametric methods of spectrum estimation is that they are not designed to incorporate information that may be available about the process into the estimation procedure. In some applications this may be an important limitation, particularly when some knowledge is available about how the data samples are generated.

Therefore, one would hope it possible to incorporate a model for the process directly into the spectrum estimation algorithm, then a more accurate and higher resolution estimate could be found. This may be easily done using a parametric approach to spectrum estimation. Although it is possible to significantly improve the resolution of the spectrum estimate with a parametric method, it is important to realize that, unless the model that is used is appropriate for the process that is being analyzed, inaccurate or misleading estimates may be obtained.

First of all, we introduce the power spectrum estimation based on the classic or nonparametric methods in section 4.2. These methods, include the periodogram, the modified periodogram, Bartlett's method, Welch's method, and the Blackman-Tukey method. And next in section 4.3, we introduce the power spectrum estimation techniques that are based on a parametric model for the data. The Models that are commonly used include autoregressive (AR), moving average (MA), autoregressive moving average (ARMA), and harmonic (complex exponentials in noise). Finally in section 4.4, we introduce some other methods of the power spectrum estimation and frequency estimation. Firstly, the minimum variance method is then considered. This technique consists of designing a narrowband filter bank to generate a set of narrowband random processes. Secondly, the maximum entropy method (MEM) is introduced and it is shown that MEM is equivalent to spectral estimation using the all-pole model. Finally we consider the frequency estimation algorithm for harmonic processes consisting of a sine wave in noise or sums of complex exponents, these methods include the Pisarenko harmonic decomposition, MUSIC, the eigenvector method, and principal components frequency estimation. In the frequency estimation algorithm, it also assumes that the process is harmonic.

4.2 Nonparametric Methods

Nonparametric techniques of spectrum estimation are considered in this section. These methods are based on the idea of estimating the autocorrelation sequence of a random process from a set of measured data, and then taking the Fourier transform to obtain an estimate of the power spectrum. Firstly, we start with the periodogram, a nonparametic method easy to compute, but it is limited in its ability to produce an accurate estimate of the power spectrum, particularly for short data records. We will then introduce a number of modifications to the periodogram that have been proposed to improve its statistical properities. These methods include the modified periodogram, Bartlett's method, Welch's method, and the Blackman-Tukey method.

4. 2. 1 The periodogram

The power spectrum of a wide-sense stationary random process can be obtained by the Fourier transform of the autocorrelation sequence,

$$P_x(e^{j\omega}) = \sum_{k=-\infty}^{\infty} r_x(k)e^{-jk\omega}$$

Therefore, spectrum estimation is an autocorrelation estimation problem. In theory, for an autocorrelation ergodic process and an unlimited amount of data, the autocorrelation sequence may be determined using the time-average

$$r_x(k) = \lim_{N\to\infty}\frac{1}{2N+1}\sum_{n=-N}^{N} x(n+k)x^*(n) \tag{4.1}$$

However, if $x(n)$ is only measured over a finite interval, $n = 0,1,\cdots N-1$, then the autocorrelation sequence must be estimated using, for example, equation (4. 1) with a finite sum,

$$\hat{r}_x(k) = \frac{1}{N}\sum_{n=0}^{N-1} x(n+k)x^*(n) \tag{4.2}$$

In order to ensure that the values of $x(n)$ that is outside the interval $[0,N-1]$ are not considered in the sum, equation (4. 2) should be rewritten

$$\hat{r}_x(k) = \frac{1}{N}\sum_{n=0}^{N-1-k} x(n+k)x^*(n)\ ; \quad k = 0,1,\cdots,N-1 \tag{4.3}$$

And the values of $\hat{r}_x(k)$ for $k<0$ are defined by using the conjugate symmetry, $\hat{r}_x(-k) = \hat{r}_x^*(k)$, with $\hat{r}_x(k)$ equal to zero for all $|k| \geqslant N$. Then the periodogram is obtained by taking the discrete-time Fourier transform of $\hat{r}_x(k)$, which leads to an estimate of the power spectrum,

$$\hat{P}_{per}(e^{j\omega}) = \sum_{k=-N+1}^{N-1} \hat{r}_x(k)e^{-jk\omega} \tag{4.4}$$

Although Eq. (4.4) is defined by considering the estimated autocorrelation sequence $\hat{r}_x(k)$, it will be easier to express this method by directly considering the process $x(n)$. Let $x_N(n)$ be the finite length signal of length N that equals $x(n)$ over the interval $[0,N-1]$, and equals zero for other values,

$$x_N(n) = \begin{cases} x(n)\ ; & 0 \leqslant n < N \\ 0\ ; & \text{otherwise} \end{cases} \tag{4.5}$$

Therefore, $x_N(n)$ can be the product of $x(n)$ with a rectangular window $w_R(n)$,

$$x_N(n) = w_R(n)x(n) \tag{4.6}$$

The estimated autocorrelation sequence may be written as follows

$$\hat{r}_x(k) = \frac{1}{N}\sum_{n=-\infty}^{\infty} x_N(n+k)x_N^*(n) = \frac{1}{N}x_N(k) * x_N^*(-k) \tag{4.7}$$

Using the convolution theorem and taking the Fourier transform, the periodogram is as follows,

$$\hat{P}_{per}(e^{j\omega}) = \frac{1}{N}X_N(e^{j\omega})X_N^*(e^{j\omega}) = \frac{1}{N}|X_N(e^{j\omega})|^2 \tag{4.8}$$

where $X_N(e^{j\omega})$ is the discrete-time Fourier transform of $x_N(n)$,

$$X_N(e^{j\omega}) = \sum_{n=-\infty}^{\infty} x_N(n)e^{-jn\omega} = \sum_{n=0}^{N-1} x(n)e^{-jn\omega} \tag{4.9}$$

Therefore, the periodogram is proportional to the squared magnitude of $X_N(e^{j\omega})$, and can be computed using a DFT as follows

$$x_N(n) \xrightarrow{\text{DFT}} X_N(k) \rightarrow \frac{1}{N}|X_N(k)|^2 = \hat{P}_{per}(e^{j2\pi k/N})$$

4.2.2 The Modified Periodogram

In Section 4.2.1, we saw that the periodogram is proportional to the squared magnitude of the Fourier transform of the windowed signal $x_N(n) = w_R(n)x(n)$,

$$\hat{P}_{per}(e^{j\omega}) = \frac{1}{N}|X_N(e^{j\omega})|^2 = \frac{1}{N}\left|\sum_{n=-\infty}^{\infty} x(n)w_R(n)e^{-jn\omega}\right|^2 \tag{4.10}$$

Instead of applying a rectangular window $w_R(n)$, equation (4.10) suggests the possibility of using other data windows. The periodogram of a process that is windowed with a general window $w(n)$ is called a modified periodogram and is given by

$$\hat{P}_M(e^{j\omega}) = \frac{1}{NU}\left|\sum_{n=-\infty}^{\infty} x(n)w(n)e^{-jn\omega}\right|^2 \tag{4.11}$$

where N is the length of the window, and U is given by

$$U = \frac{1}{N}\sum_{n=0}^{N-1}|w(n)|^2 \tag{4.12}$$

where U is a constant that, as we will see, is defined so that $\hat{P}_M(e^{j\omega})$ will be asymptotically unbiased. Moreover, the amount of smoothing in the periodogram is determined by the window that is applied to the data. Although a rectangular window has a narrow main lobe compared to other windows, it has relatively larger side lobes that may lead to masking of weak narrowband components. In other words, this reduction in the side lobes amplitude by using other windows comes at the expense of an increase in the width of the main lobe, which affects the resolution in turn.

4.2.3 Other Nonparametric Methods

Bartlett's Method: Periodogram Averaging. Bartlett's method of periodogram averaging which unlike either the periodogram or the modified periodogram, produces a consistent estimate of the power spectrum. The motivation for this method comes from the observation that the expected value of the periodogram converges to $P_x(e^{j\omega})$ with the data record length N going to infinity,

$$\lim_{N\to\infty} E\{\hat{P}_{per}(e^{j\omega})\} = P_x(e^{j\omega}) \tag{4.13}$$

Thus, the estimate will be a consistent estimate of $P_x(e^{j\omega})$, if we can find a consistent estimate of the mean, $E\{\hat{P}_{per}(e^{j\omega})\}$.

Let $x_i(n)$ for $i = 1,2,\cdots,K$ be K uncorrelated realizations of a random process $x(n)$ over the interval $0 \leqslant n < L$. The periodogram of $x_i(n)$ as follows:

$$\hat{P}_{\text{per}}^{(i)}(\mathrm{e}^{\mathrm{j}\omega}) = \frac{1}{L}\left|\sum_{n=0}^{L-1} x_i(n)\mathrm{e}^{-\mathrm{j}n\omega}\right|^2; \quad i = 1,2,\cdots,K \tag{4.14}$$

The average of these periodogram is

$$\hat{P}_x(\mathrm{e}^{\mathrm{j}\omega}) = \frac{1}{K}\sum_{i=1}^{K}\hat{P}_{\text{per}}^{(i)}(\mathrm{e}^{\mathrm{j}\omega}) \tag{4.15}$$

And we have the expectation of $\hat{P}_x(\mathrm{e}^{\mathrm{j}\omega})$ as follows:

$$E\{\hat{P}_x(\mathrm{e}^{\mathrm{j}\omega})\} = E\{\hat{P}_{\text{per}}^{(i)}(\mathrm{e}^{\mathrm{j}\omega})\} = \frac{1}{2\pi}P_x(\mathrm{e}^{\mathrm{j}\omega}) * W_B(\mathrm{e}^{\mathrm{j}\omega}) \tag{4.16}$$

where $W_B(\mathrm{e}^{\mathrm{j}\omega})$ is the Fourier transform of Bartlett window, which extends from $-L$ to L. The $w_B(k)$ and $W_B(\mathrm{e}^{\mathrm{j}\omega})$ are defined as follows:

$$w_B(k) = \begin{cases} \dfrac{N-|k|}{N}; & |k| \leqslant N \\ 0; & |k| > N \end{cases} \tag{4.17}$$

$$W_B(\mathrm{e}^{\mathrm{j}\omega}) = \frac{1}{N}\left[\frac{\sin(N\omega/2)}{\sin(\omega/2)}\right]^2 \tag{4.18}$$

Therefore, as with the periodogram, $\hat{P}_x(\mathrm{e}^{\mathrm{j}\omega})$ is asymptotically unbiased. However, the hardship with this approach is that uncorrelated realizations of a process are generally not available. Instead, one typically only has a single realization of length N. Therefore, Bartlett proposed that $x(n)$ be partitioned into K nonoverlapping sequences of length L, where $N = KL$. The Bartlett estimate is then computed as in equation (4.14) and equation (4.15) with

$$x_i(n) = x(n+iL) \quad n = 0,1,\cdots,L-1$$
$$i = 0,1,\cdots,K-1$$

Therefore, the Bartlett estimate is

$$\hat{P}_B(\mathrm{e}^{\mathrm{j}\omega}) = \frac{1}{N}\sum_{i=0}^{K-1}\left|\sum_{n=0}^{L-1} x(n+iL)\mathrm{e}^{-\mathrm{j}n\omega}\right|^2 \tag{4.19}$$

As in equation (4.16), the expected value of Bartlett's estimate is

$$E\{\hat{P}_B(\mathrm{e}^{\mathrm{j}\omega})\} = \frac{1}{2\pi}P_x(\mathrm{e}^{\mathrm{j}\omega}) * W_B(\mathrm{e}^{\mathrm{j}\omega}) \tag{4.20}$$

Welch's Method: Averaging Modified Periodograms. Welch proposed two modifications to Bartlett's method. The first modification is to allow the $x_i(n)$ to overlap, and the second is to allow a data window $w(n)$ to be applied to each sequence, thereby producing a set of modified periodograms that are to be averaged.

Assuming that successive sequences are offset by D values and that each sequence is L values long then the i th sequence is as follows

$$x_i(n) = x(n+iD); \quad n = 0,1,\cdots,L-1 \tag{4.21}$$

As a result, the overlap between $x_i(n)$ and $x_{i+1}(n)$ is $L-D$ values, and if K sequences cover the entire N data values, then we have

$$N = L + D(K - 1) \tag{4.22}$$

with no overlap ($D=L$), we have $K = N/L$ parts of length L which leads to Bartlett's method. If the sequences are allowed to overlap by 50% ($D=L/2$), then we may obtain

$$K = 2\frac{N}{L} - 1 \tag{4.23}$$

parts of length L, that maintaining the same part length as Bartlett's method while doubling the number of modified periodograms to be averaged. Therefore, by allowing the sequences to overlap, it is possible to increase the number and/or length of the sequences that are averaged, thereby weakening the reduction in resolution.

Blackman-Tukey Method: Periodogram Smoothing. Bartlett's and Welch's methods are designed to reduce the variance of the periodogram by averaging periodograms and modified periodograms. Another method for decreasing the statistical variability of the periodogram is periodogram smoothing, which is often referred to as the Blackman-Tukey method.

In the Blackman-Tukey method, the variance of the periodogram is reduced by applying a window to $\hat{r}_x(k)$ in order to decrease the contribution of the unreliable estimates to the periodogram. Specifically, the Blackman-Tukey spectrum estimate is

$$\hat{P}_{\mathrm{BT}}(\mathrm{e}^{\mathrm{j}\omega}) = \sum_{k=-M}^{M} \hat{r}_x(k) w(k) \mathrm{e}^{-\mathrm{j}k\omega} \tag{4.24}$$

where $w(k)$ is a *lag window* that is applied to the autocorrelation estimate. For example, if $w(k)$ is a rectangular window extending from $-M$ to M with $M < N - 1$, then the estimates of $r(k)$ having the largest variance are set to zero and the power spectrum estimate will have a smaller variance. What is traded for this reduction in variance is a reduction in resolution since a smaller number of autocorrelation estimates are used to form the estimate of the power spectrum.

Using the frequency convolution theorem, the Blackman-Tukey spectrum may be written in the frequency domain as follows:

$$\hat{P}_{\mathrm{BT}}(\mathrm{e}^{\mathrm{j}\omega}) = \frac{1}{2\pi}\hat{P}_{\mathrm{per}}(\mathrm{e}^{\mathrm{j}\omega}) * W(\mathrm{e}^{\mathrm{j}\omega}) = \frac{1}{2\pi}\int_{-\pi}^{\pi} \hat{P}_{\mathrm{per}}(\mathrm{e}^{\mathrm{j}u}) W(\mathrm{e}^{\mathrm{j}(\omega-u)})\, du \tag{4.25}$$

Thus, the Blackman-Tukey method smooths the periodogram by convolving with the Fourier transform of the autocorrelation window $W(\mathrm{e}^{\mathrm{j}\omega})$. Although there is a certain amount of flexibility in the choice of the window to be used, it should be conjugate symmetric so that $W(\mathrm{e}^{\mathrm{j}\omega})$ is real-valued, and the window should have a non-negative Fourier transform $W(\mathrm{e}^{\mathrm{j}\omega}) \geqslant 0$ so that $\hat{P}_{\mathrm{BT}}(\mathrm{e}^{\mathrm{j}\omega})$ is guaranteed to be non-negative.

4.3 Parametric Methods

Nonparametric methods of spectrum estimation presented in sections before are not designed to incorporate information that may be available about the process into the estimation procedure. This may be a significant limitation in some applications, particularly when some knowledge is available about how the data samples are generated.

Therefore, if it is possible to incorporate a model for the process directly into the spectrum estimation, then a more accurate and higher resolution estimate could be obtained. This may be easily done by using a parametric approach to spectrum estimation.

The steps of parametric approach are as follows:

1. Select an appropriate model for the process, which could be based on a priori knowledge about how the process is generated, or experimental results indicating that a particular model works well.

2. Models that are commonly used include autoregressive (AR), moving average (MA), autoregressive moving average (ARMA), and harmonic (complex exponentials in noise).

3. Once a model has been selected, the next step is to estimate the model parameters from the given data.

4. The final step is to estimate the power spectrum by incorporating the estimated parameters into the parametric form for the spectrum.

4.3.1 Autoregressive Spectrum Estimation

An autoregressive process, $x(n)$, may be represented as the output of an all-pole filter that is driven by unit variance white noise. The power spectrum of a pth-order autoregressive process is

$$P_x(e^{j\omega}) = \frac{|b(0)|^2}{\left|1 + \sum_{k=1}^{p} a_p(k) e^{-jk\omega}\right|^2} \tag{4.26}$$

Therefore, if $b(0)$ and $a_p(k)$ can be estimated from the data, then an estimate of the power spectrum may be formed using

$$\hat{P}_{AR}(e^{j\omega}) = \frac{|\hat{b}(0)|^2}{\left|1 + \sum_{k=1}^{p} \hat{a}_p(k) e^{-jk\omega}\right|^2} \tag{4.27}$$

Clearly, the accuracy of $P_{AR}(e^{j\omega})$ will depend on how accurately the model parameters may be estimated and whether or not an autoregressive model is consistent with the way in which the data is generated.

Since autoregressive spectrum estimation requires that an all-pole model be found for the process, there is a variety of techniques that may be used to estimate the all-pole parameters. However, once the all-pole parameters have been estimated, each method generates an estimate of the power spectrum in exactly the same way. In the following subsections, we briefly review the AR modeling techniques and describe some of the properties of these techniques as they are applied to spectrum estimation.

The Autocorrelation Method. In the autocorrelation method of all-pole modeling, the AR coefficients $a_p(k)$ are found by solving the autocorrelation normal equations.

$$\begin{bmatrix} r_x(0) & r_x^*(1) & r_x^*(2) & \cdots & r_x^*(p) \\ r_x(1) & r_x^*(0) & r_x^*(1) & \cdots & r_x^*(p-1) \\ r_x(2) & r_x^*(1) & r_x^*(0) & \cdots & r_x^*(p-2) \\ \vdots & \vdots & \vdots & \ddots & \vdots \\ r_x(p) & r_x^*(p-1) & r_x^*(p-2) & \cdots & r_x^*(0) \end{bmatrix} \begin{bmatrix} 1 \\ a_p(1) \\ a_p(2) \\ \vdots \\ a_p(p) \end{bmatrix} = \varepsilon_p \begin{bmatrix} 1 \\ 0 \\ 0 \\ \vdots \\ 0 \end{bmatrix} \tag{4.28}$$

where

$$r_x(k) = \frac{1}{N} \sum_{n=0}^{N-1-k} x(n+k)x^*(n), k = 0,1,\cdots,p \tag{4.29}$$

(for simplicity we are suppressing the hats over the autocorrelations $r_x(k)$ and all-pole parameters $a_p(k)$). Solving the above equation for the coefficients $a_p(k)$, setting

$$|b(0)|^2 = \varepsilon_p = r_x(0) + \sum_{k=1}^{p} a_p(k) r_x^*(k) \tag{4.30}$$

and incorporating these parameters into equation (4.27) produces an estimate of the power spectrum, which is sometimes referred to as the Yule-Walker method. Note that the Yule-Walker method is equivalent to the maximum entropy method. In fact, the only difference between the two methods lies in the assumptions that are made about the process $x(n)$. Specifically, it is assumed that $x(n)$ is an autoregressive process in the Yule–Walker method, whereas it is assumed that $x(n)$ is Gaussian in the maximum entropy method.

The Covariance Method. Another approach for estimating the AR parameters is the covariance method. The covariance method requires finding the solution to the set of linear equations,

$$\begin{bmatrix} r_x(1,1) & r_x(2,1) & \cdots & r_x(p,1) \\ r_x(1,2) & r_x(2,2) & \cdots & r_x(p,2) \\ \vdots & \vdots & \ddots & \vdots \\ r_x(1,p) & r_x(2,p) & \cdots & r_x(p,p) \end{bmatrix} \begin{bmatrix} a_p(1) \\ a_p(2) \\ \vdots \\ a_p(p) \end{bmatrix} = - \begin{bmatrix} r_x(0,1) \\ r_x(0,2) \\ \vdots \\ r_x(0,p) \end{bmatrix} \tag{4.31}$$

where

$$r_x(k,l) = \sum_{n=p}^{N-1} x(n-l)x^*(n-k) \tag{4.32}$$

Unlike the linear equations in the autocorrelation method, these equations are not Toeplitz. However, the advantage of the covariance method over the autocorrelation method is that no windowing of the data is required in the formation of the autocorrelation estimates, $r_x(k,l)$. Therefore, for short data records the covariance method generally produces higher resolution spectrum estimates than the autocorrelation method. However, as the data record length increases and becomes large compared to the model order, $N \gg p$, the effect of the data window becomes small and the difference between the two approaches becomes negligible.

The Modified Covariance Method. The modified covariance method is similar to the covariance method in that no window is applied to the data. However, instead of finding the autoregressive model that minimizes the sum of the squares of the forward prediction error, the modified covariance method minimizes the sum of the squares of the forward and backward prediction errors. As a result, the autoregressive parameters in the modified covariance method are found by solving a set of linear equations of the form given before with

$$r_x(k,l) = \sum_{n=p}^{N-1} [x(n-l)x^*(n-k) + x(n-p+l)x^*(n-p+k)] \qquad (4.33)$$

replacing the estimate in the Covariance Method. Just like in the covariance method, the autocorrelation matrix is not Toeplitz.

In contrast to other AR spectrum estimation techniques, the modified covariance method appears to give statistically stable spectrum estimates with high resolution. Furthermore, in the spectral analysis of sinusoids in white noise, although the modified covariance method is characterized by a shifting of the peaks from their true locations due to additive noise, this shifting appears to be less pronounced than in other autoregressive estimation techniques. In addition, the peak locations tend to be less sensitive to phase. Finally, unlike the previous methods, it appears that the modified covariance method is not subject to spectral line splitting.

Selecting the Model Order. A question that remains to be answered in the use of an AR spectrum estimation method is how to select the model order p of the AR process. If the model order used is too small, then the resulting spectrum will be smoothed and will have poor resolution. If, on the other hand, the model order is too large, then the spectrum may contain spurious peaks and may lead to spectral line splitting. Therefore, it would be useful to have a criterion that indicates the appropriate model order to use for a given set of data. One approach would be to increase the model order until the modeling error is minimized. However, the difficulty in this approach is that the error is a monotonically nonincreasing function of the model order p. This problem may be overcome by incorporating a penalty function that increases with the model order p. Several criteria have been proposed including a penalty term that increases linearly with p,

$$C(p) = N\log\varepsilon_p + f(N)p \qquad (4.34)$$

Here, ε_p is the modeling error, N is the data record length, and $f(N)$ is a constant that may depend upon N. The idea, then, is to select the value of p that minimizes $C(p)$. Two criteria that are of this form are the Akaike Information Criterion.

$$\mathrm{AIC}(p) = N\log\varepsilon_p + 2p \qquad (4.35)$$

and the minimum description length proposed by Rissanen,

$$\mathrm{MDL}(p) = N\log\varepsilon_p + (\log N)p \qquad (4.36)$$

The AIC was derived by minimizing an information theoretic function and includes the penalty $2p$ for any extra AR coefficients that do not significantly reduce the prediction error. It has been observed that the AIC gives an estimate for the order p that is too small when applied to non-autoregressive processes and that it tends to overestimate the order as N increases. The MDL, on the other hand, contains the penalty term $(\log N)p$, which increases with the data record length N and the model order p. It has been shown that the MDL is a consistent model-order estimator in the sense that it converges to the true order as the number of observations N increases. Two other model order selection criteria that are often used are Akaike's Final Prediction Error,

$$\mathrm{FPE}(p) = \varepsilon_p \frac{N+p+1}{N-p-1} \qquad (4.37)$$

and Parzen's Criterion Autoregressive Transfer function

$$\mathrm{CAT}(p) = \left[\frac{1}{N}\sum_{j=1}^{p}\frac{N-j}{N\varepsilon_j}\right] - \frac{N-p}{N\varepsilon_p} \tag{4.38}$$

4.3.2 Moving Average Spectrum Estimation

A moving average process may be generated by filtering unit variance white noise, $w(n)$, with an FIR filter as follows:

$$x(n) = \sum_{k=0}^{q} b_q(k) w(n-k) \tag{4.39}$$

the relationship between the power spectrum of a moving average process and the coefficients $b_q(k)$ is

$$P_x(e^{j\omega}) = \left|\sum_{k=0}^{q} b_q(k) e^{-jk\omega}\right|^2 \tag{4.40}$$

Equivalently, the power spectrum may be written in terms of the autocorrelation sequence $r_x(k)$ as

$$P_x(e^{j\omega}) = \sum_{k=-q}^{q} r_x(k) e^{-jk\omega} \tag{4.41}$$

where $r_x(k)$ is related to the filter coefficients $b_q(k)$ through the Yule-Walker equations

$$r_x(k) = \sum_{l=0}^{q-k} b_q(n+l) b_q^*(l) \quad , \quad k = 0,1,\cdots,p \tag{4.42}$$

with $r_x(-k) = r_x^*(k)$ and $r_x(k) = 0$ for $|k| > q$.

In a moving average model, the spectrum may be estimated in two ways. The first approach is to take advantage of the fact that the autocorrelation sequence of a moving average process is finite in length. Specifically, since $r_x(k) = 0$ for $|k| > q$, then a natural estimate to use is

$$\hat{P}_{\mathrm{MA}}(e^{j\omega}) = \sum_{k=-q}^{q} \hat{r}_x(k) e^{-jk\omega} \tag{4.43}$$

where $\hat{r}_x(k)$ is a suitable estimate of the autocorrelation sequence. Note that although $\hat{P}_{\mathrm{MA}}(e^{j\omega})$ is equivalent to the Blackman-Tukey estimate (a kind of non-parametric method) using a rectangular window, there is a subtle difference in the assumptions that are behind these two estimates. In particular, since it assumes that $x(n)$ is a moving average process of order q, then the true autocorrelation sequence is zero for $|k| > q$. Thus, if an unbiased estimate of the autocorrelation sequence is used for $|k| \leqslant q$, then

$$E\{\hat{P}_{\mathrm{MA}}(e^{j\omega})\} = P_x(e^{j\omega}) \tag{4.44}$$

namely, $\hat{P}_{\mathrm{MA}}(e^{j\omega})$ is unbiased. The Blackman-Tukey method, on the other hand, makes no assumptions about $x(n)$ and may be applied to any type of process. Therefore, due to the windowing of the autocorrelation sequence, unless $x(n)$ is a moving average process, the Blackman-Tukey spectrum will be biased.

The second approach is to estimate the moving average parameters, $b_q(k)$, from $x(n)$ and then substitute these estimates as follows:

$$\hat{P}_{MA}(e^{j\omega}) = \left| \sum_{k=0}^{q} \hat{b}_q(k) e^{-jk\omega} \right|^2 \tag{4.45}$$

Like in the autoregressive spectrum estimation, it is useful to have a criterion for estimating the order of the MA model that should be used for a given process $x(n)$.

4.3.3 Autoregressive Moving Average Spectrum Estimation

An autoregressive moving average process has a power spectrum of the form

$$P_x(e^{j\omega}) = \frac{\left| \sum_{k=0}^{q} b_q(k) e^{-jk\omega} \right|^2}{\left| 1 + \sum_{k=1}^{p} a_p(k) e^{-jk\omega} \right|^2} \tag{4.46}$$

which may be generated by filtering unit variance white noise with a filter having both poles and zeros,

$$H(z) = \frac{B_q(z)}{A_P(z)} = \frac{\sum_{k=0}^{q} b_q(k) z^{-k}}{1 + \sum_{k=1}^{p} a_p(k) z^{-k}} \tag{4.47}$$

Following the approach used for AR(p) and MA(q) spectrum estimation, the spectrum of an ARMA (p,q) process may be estimated using estimates of the model parameters

$$\hat{P}_{ARMA}(e^{j\omega}) = \frac{\left| \sum_{k=0}^{q} \hat{b}_q(k) e^{-jk\omega} \right|^2}{\left| 1 + \sum_{k=1}^{p} \hat{a}_p(k) e^{-jk\omega} \right|^2} \tag{4.48}$$

As we discussed before, the AR model parameters may be estimated from the modified Yule-Walker equations either directly or by using a Least Squares approach. Once the coefficients $\hat{a}_q(k)$ have been estimated, a moving average modeling technique such as Durbin's method may be used to estimate the moving average parameters $\hat{b}_q(k)$.

4.4 Other Methods

In this section, we introduce some other methods of spectrum estimation, including the Minimum Variance (MV) method, the Maximum Entropy method, the Frequency estimation, and the Principal Components Spectrum Estimation.

4.4.1 Minimum Variance Spectrum Estimation

In this section, we develop the Minimum Variance (MV) method of spectrum estimation, which is an adaptation of the Maximum Likelihood Method (MLM) developed for the analysis of two-dimensional power spectral densities. In the MV method, the power spectrum is estimated by filtering a WSS random process with a narrowband bandpass filter. Therefore, let $x(n)$ be a zero mean WSS random process with a power spectrum $P_x(e^{j\omega})$ and define $g_i(n)$ as an ideal bandpass filter with the center frequency ω_i and a bandwidth Δ,

$$|G_i(e^{j\omega})| = \begin{cases} 1; |\omega - \omega_i| \leq \Delta/2 \\ 0; \text{otherwise} \end{cases} \tag{4.49}$$

And the power spectrum of the output process $y_i(n)$ is

$$P_i(e^{j\omega}) = P_x(e^{j\omega}) \, |G_i(e^{j\omega})|^2 \tag{4.50}$$

And the power in $y_i(n)$ is

$$\begin{aligned} E\{|y_i(n)|^2\} &= \frac{1}{2\pi}\int_{-\pi}^{\pi} P_i(e^{j\omega})\,d\omega = \frac{1}{2\pi}\int_{-\pi}^{\pi} P_x(e^{j\omega})\,|G_i(e^{j\omega})|^2 d\omega \\ &= \frac{1}{2\pi}\int_{\omega_i-\Delta/2}^{\omega_i+\Delta/2} P_x(e^{j\omega})\,d\omega \end{aligned} \tag{4.51}$$

If Δ is small enough and $P_x(e^{j\omega})$ is approximately constant over the passband of the filter, then power in $y_i(n)$ is as follows:

$$E\{|y_i(n)|^2\} \approx P_x(e^{j\omega_i}) \frac{\Delta}{2\pi} \tag{4.52}$$

Then, it is possible to estimate the power spectral density of $x(n)$ at $\omega = \omega_i$ from the filtered process by estimating the power in $y_i(n)$, and dividing by the normalized filter bandwidth $\frac{\Delta}{2\pi}$,

$$\hat{P}_x(e^{j\omega_i}) = \frac{E\{|y_i(n)|^2\}}{\Delta/2\pi} \tag{4.53}$$

Since each filter in the filter bank for the periodogram is the same, differing only in the center frequency, these filters are data-independent. As a result of that, when a random process contains a large amount of power in frequency bands within the side lobes of the bandpass filter, leakage through the side lobes may lead to significant distortion in the power estimates.

Therefore, a better approach should allow each filter in the filter bank to be *data-adaptive* so that each filter may be designed to be "optimum" in the sense of rejecting as much out-of-band signal power as possible.

The minimum variance spectrum estimation technique is based on this idea and involves the following steps:

1. Design a bank of bandpass filters $g_i(n)$ with center frequency and each filter rejects the maximum amount of out-of-band power when passing the component at frequency ω_i with no distortion.

2. Filter $x(n)$ using each filter in the filter bank and estimate the power in each output process $y_i(n)$.

3. Let $\hat{P}_x(e^{j\omega_i})$ equal to the result of power estimated in step (2) divided by the filter bandwidth.

Next, to derive the minimum variance spectrum estimate, we design the bandpass filter bank first.

Let $g_i(n)$ be a complex-valued FIR bandpass filter of order p. To ensure that the filter does not change the power in the input process at frequency ω_i, $G_i(e^{j\omega})$ will be constrained to have a gain of one at $\omega = \omega_i$,

$$G_i(e^{j\omega_i}) = \sum_{n=0}^{p} g_i(n) e^{-jn\omega_i} = 1 \tag{4.54}$$

Let $\boldsymbol{g}_i$ and $\boldsymbol{e}_i$ be the vector filter coefficients $g_i(n)$ and the vector of complex exponentials $e^{jk\omega_i}$ respectively,

$$\boldsymbol{g}_i = [g_i(0), g_i(1), \cdots, g_i(p)]^T$$

$$\boldsymbol{e}_i = [1, e^{j\omega_i}, \cdots, e^{jp\omega_i}]^T$$

Then equation (4.54) can be written in vector form:

$$\boldsymbol{g}_i^H \boldsymbol{e}_i = \boldsymbol{e}_i^H \boldsymbol{g}_i = 1 \tag{4.55}$$

Since each filter should reject the maximum amount of out-of-band power, the bandpass filter to be designed should minimize the power in the output process subject to the linear constraint given in equation (4.55). Through previous knowledge of discrete-time random processes, the power in $y_i(n)$ can be expressed in terms of the autocorrelation matrix $\boldsymbol{R}_x$:

$$E\{|y_i(n)|^2\} = \boldsymbol{g}_i^H \boldsymbol{R}_x \boldsymbol{g}_i \tag{4.56}$$

The solution to this problem is

$$\boldsymbol{g}_i = \frac{\boldsymbol{R}_x^{-1} \boldsymbol{e}_i}{\boldsymbol{e}_i^H \boldsymbol{R}_x^{-1} \boldsymbol{e}_i} \tag{4.57}$$

And the minimum value of $E\{|y_i(n)|^2\}$ is equal to

$$\min_{g_i} E\{|y_i(n)|^2\} = \frac{1}{\boldsymbol{e}_i^H \boldsymbol{R}_x^{-1} \boldsymbol{e}_i} \tag{4.58}$$

Note that although these equations were derived for a specific frequency ω_i, since this frequency was arbitrary, then these equations are valid for all ω. Thus, the optimum filter at frequency ω is

$$\boldsymbol{g} = \frac{\boldsymbol{R}_x^{-1} \boldsymbol{e}}{\boldsymbol{e}^H \boldsymbol{R}_x^{-1} \boldsymbol{e}} \tag{4.59}$$

And we have

$$\hat{P}_x(e^{j\omega_i}) = \frac{E\{|y_i(n)|^2\}}{\Delta/2\pi} = \frac{1}{\boldsymbol{e}^H \boldsymbol{R}_x^{-1} \boldsymbol{e}} \frac{2\pi}{\Delta} \tag{4.60}$$

where $\boldsymbol{e} = [1, e^{j\omega}, \cdots, e^{jp\omega}]^T$. If we set $\Delta = \frac{2\pi}{p+1}$, then in general, for all ω we have:

$$\hat{P}_{MV}(e^{j\omega}) = \frac{p+1}{\boldsymbol{e}^H \boldsymbol{R}_x^{-1} \boldsymbol{e}} \tag{4.61}$$

which is the *minimum variance spectrum estimate*. Note that $\hat{P}_{MV}(e^{j\omega})$ is defined in terms of the autocorrelation matrix $\boldsymbol{R}_x$ of $x(n)$. Normally, the $\boldsymbol{R}_x$ is unknown, then it can be replaced with an estimation $\hat{\boldsymbol{R}}_x$.

4.4.2 Maximum Entropy Method

Given the autocorrelation $r_x(k)$ of a WSS process for lags $|k| \leqslant p$, the problem that we wish to address is how to extrapolate $r_x(k)$ for $|k| > p$. Denoting the extrapolated values by $r_e(k)$, it is clear that some constraints should be placed on $r_e(k)$. For example,

$$P_x(e^{j\omega}) = \sum_{k=-p}^{p} r_x(k) e^{-jk\omega} + \sum_{|k|>p} r_e(k) e^{-jk\omega} \tag{4.62}$$

In general, only constraining $P_x(e^{j\omega})$ to be real and non-negative is not sufficient to guarantee the extrapolation. Therefore, some additional constraints must be imposed on the set of allowable extrapolations. One such constraint is to maximize the entropy of the process, since the entropy is a measure of randomness or uncertainty. In terms of the power spectrum, this corresponds to the constraint that $P_x(e^{j\omega})$ be as flat as possible (with less fluctuation).

For a Gaussian random process with power spectrum $P_x(e^{j\omega})$, the entropy is

$$H(x) = \frac{1}{2\pi}\int_{-\pi}^{\pi} \ln P_x(e^{j\omega}) d\omega \tag{4.63}$$

Therefore, for Gaussian processes with a given partial autocorrelation sequence $r_x(k)$ for $|k| \leqslant p$, the maximum entropy power spectrum is the one that maximizes $H(x)$ subject to the constraint that the inverse discrete-time Fourier transform of $P_x(e^{j\omega})$ equals the given set of autocorrelations for $|k| \leqslant p$,

$$r_x(k) = \frac{1}{2\pi}\int_{-\pi}^{\pi} P_x(e^{j\omega}) e^{jk\omega} d\omega;\ |k| \leqslant p \tag{4.64}$$

The values of $r_e(k)$ that maximize the entropy may be found by setting the derivative of $H(x)$ with respect to $r_e^*(k)$ equal to zero as follows.

$$\frac{\partial H(x)}{\partial r_e^*(k)} = \frac{1}{2\pi}\int_{-\pi}^{\pi} \frac{1}{P_x(e^{j\omega})} \frac{\partial P_x(e^{j\omega})}{\partial r_e^*(k)} d\omega = 0;\ |k| > p \tag{4.65}$$

From equation (4.62) and equation (4.65) we can obtain that (recall that $r_x(-k) = r_x^*(k)$):

$$\frac{\partial P_x(e^{j\omega})}{\partial r_e^*(k)} = e^{jk\omega} \tag{4.66}$$

$$\frac{1}{2\pi}\int_{-\pi}^{\pi} \frac{1}{P_x(e^{j\omega})} e^{jk\omega} d\omega = 0;\ |k| > p \tag{4.67}$$

Defining $Q_x(e^{j\omega}) = \dfrac{1}{P_x(e^{j\omega})}$, the inverse discrete-time Fourier transform of $Q_x(e^{j\omega})$ is a finite-length sequence that is equal to zero for $|k| > p$.

$$q_x(k) = \frac{1}{2\pi}\int_{-\pi}^{\pi} Q_x(e^{j\omega}) e^{jk\omega} d\omega = 0;\ |k| > p \tag{4.68}$$

Therefore,

$$Q_x(e^{j\omega}) = \frac{1}{P_x(e^{j\omega})} = \sum_{k=-p}^{p} q_x(k) e^{-jk\omega} \tag{4.69}$$

And it follows that the maximum entropy power spectrum for a Gaussian process, which we will denote by $\hat{P}_{\text{mem}}(e^{j\omega})$, is an all-pole power spectrum,

$$\hat{P}_{\text{mem}}(e^{j\omega}) = \frac{1}{\sum_{k=-p}^{p} q_x(k) e^{-jk\omega}} \tag{4.70}$$

Using the spectral factorization theorem, it can be expressed as

$$\hat{P}_{\text{mem}}(e^{j\omega}) = \frac{|b(0)|^2}{A_p(e^{j\omega})A_p^*(e^{j\omega})} = \frac{|b(0)|^2}{\left|1 + \sum_{k=1}^{p} a_p(k)e^{-jk\omega}\right|^2} \tag{4.71}$$

In terms of the vectors $\boldsymbol{a}_p = [1 \quad a_p(1) \quad \cdots \quad a_p(p)]^{\text{T}}$ and $\boldsymbol{e} = [1 \quad e^{j\omega} \quad \cdots \quad e^{jp\omega}]^{\text{T}}$, the MEM spectrum may be written as

$$\hat{P}_{\text{mem}}(e^{j\omega}) = \frac{|b(0)|^2}{|\boldsymbol{e}^{\text{H}} \boldsymbol{a}_p|^2} \tag{4.72}$$

Then we need to find the coefficients $a_p(k)$ and $b(0)$. Due to the constraint in equation (4.64), these coefficients must be chosen in such a way that the inverse discrete-time Fourier transform of $\hat{P}_{\text{mem}}(e^{j\omega})$ produces an autocorrelation sequence that matches the given values of $r_x(k)$ for $|k| \leqslant p$. If the coefficients $a_p(k)$ are the solution to the autocorrelation normal equations (as we can see in chapter 2),

$$\begin{bmatrix} r_x(0) & r_x^*(1) & \cdots & r_x^*(p) \\ r_x(1) & r_x(0) & \cdots & r_x^*(p-1) \\ \vdots & \vdots & \ddots & \vdots \\ r_x(p) & r_x(p-1) & \cdots & r_x(0) \end{bmatrix} \begin{bmatrix} 1 \\ a_p(1) \\ \vdots \\ a_p(p) \end{bmatrix} = \varepsilon_p \begin{bmatrix} 1 \\ 0 \\ \vdots \\ 0 \end{bmatrix} \tag{4.73}$$

and if

$$|b(0)|^2 = r_x(0) + \sum_{k=1}^{p} a_p(k) r_x^*(k) = \varepsilon_p \tag{4.74}$$

then the autocorrelation matching constraint given in equation (4.64) will be satisfied. Thus, the MEM spectrum is

$$\hat{P}_{\text{mem}}(e^{j\omega}) = \frac{\varepsilon_p}{|\boldsymbol{e}^{\text{H}} \boldsymbol{a}_p|^2} \tag{4.75}$$

In summary, given a sequence of autocorrelations, $r_x(k)$ for $k = 0, 1, \cdots, p$, the MEM spectrum is computed as follows. First, the autocorrelation normal equations (4.73) are solved for the all-pole coefficients $a_p(k)$ and ε_p. Then, the MEM spectrum is formed by incorporating these parameters into equation (4.75).

The properties of the maximum entropy method have been studied extensively and, as a spectrum analysis tool, the maximum entropy method is subject to different interpretations. It may be argued that in the absence of any information or constraints on a process, given a set of autocorrelation values, the best way to estimate the power spectrum is to Fourier Transform the autocorrelation sequence formed from the given values along with an extrapolation that imposes the least amount of structure on the data, which is a maximum entropy extrapolation. On the other hand, it may also be argued that since the maximum entropy extrapolation imposes an all-pole model on the data, unless the process is known to be consistent with this model, then the estimated spectrum may not be very accurate.

4.4.3 Frequency Estimation

As we can see in previous sections, the power spectrum of a WSS random process could be modeled as the output of a linear shift-invariant filter that is driven by white noise.

Another significant model is to consider $x(n)$ as a sum of complex exponentials in white noise,

$$x(n) = \sum_{i=1}^{p} A_i e^{jn\omega_i} + w(n) \tag{4.76}$$

The amplitudes A_i are assumed to be complex,

$$A_i = |A_i| e^{j\phi_i}$$

where ϕ_i uncorrelated random variables are uniformly distributed over the interval $[-\pi, \pi]$. Although the frequencies ω_i and magnitudes $|A_i|$ of the complex exponentialls are not random, they are assumed to be unknown. Therefore, the power spectrum of $x(n)$ consists of a set of p impulses at frequency ω_i for $i = 1, 2, \cdots, p$, and the power spectrum of the additive noise $w(n)$.

Signals of this form are found in a number of applications such as sonar signal processing and speech processing. Besides, the complex exponentials are the "information containing" part of the signal, and it is the estimation of the frequencies and amplitudes that is of interest, rather than the estimation of the power spectrum itself. For example, in the case of sonar signals, the frequencies ω_i may represent the velocity information.

In this section, we consider *frequency estimation* algorithms that consider the known properties of the process. These methods are based on an eigen-decomposition of the autocorrelation matrix into a signal subspace and a noise subspace. Once the two subspaces determined, a frequency estimation function is used to extract estimates of the frequencies.

Eigen-decomposition of the autocorrelation matrix. Consider the first-order process that consists of a single complex exponential in white noise,

$$x(n) = A_1 e^{jn\omega_1} + w(n)$$

where the amplitude of the complex exponential is $A_i = |A_i| e^{j\phi_i}$ in which ϕ_1 is a uniformly distributed random variable, and the $w(n)$ is white noise with a variance σ_w^2. The autocorrelation sequence of $x(n)$ is

$$r(k) = P_1 e^{jk\omega_1} + \sigma_w^2 \delta(k)$$

where $P_1 = |A_1|^2$ is the power in complex exponential. Thus, the $M \times M$ autocorrelation matrix for $x(n)$ is the sum of the signal autocorrelation matrix $\boldsymbol{R}_s$ and the noise autocorrelation matrix $\boldsymbol{R}_n$,

$$\boldsymbol{R}_x = \boldsymbol{R}_s + \boldsymbol{R}_n \tag{4.77}$$

where the signal autocorrelation matrix with a rank of one is as follows:

$$\boldsymbol{R}_s = P_1 \begin{bmatrix} 1 & e^{-j\omega_1} & e^{-j2\omega_1} & \cdots & e^{-j(M-1)\omega_1} \\ e^{j\omega_1} & 1 & e^{-j\omega_1} & \cdots & e^{-j(M-2)\omega_1} \\ e^{j2\omega_1} & e^{j\omega_1} & 1 & \cdots & e^{-j(M-3)\omega_1} \\ \vdots & \vdots & \vdots & \ddots & \vdots \\ e^{j(M-1)\omega_1} & e^{j(M-2)\omega_1} & e^{j(M-3)\omega_1} & \cdots & 1 \end{bmatrix} \tag{4.78}$$

And the noise autocorrelation matrix with a full rank is as follows:

$$\boldsymbol{R}_n=\sigma_w^2\boldsymbol{I}$$

If we define

$$\boldsymbol{e}_1=[1,\mathrm{e}^{\mathrm{j}\omega_1},\mathrm{e}^{\mathrm{j}2\omega_1},\cdots,\mathrm{e}^{\mathrm{j}(M-1)\omega_1}]^{\mathrm{T}} \tag{4.79}$$

we have

$$\boldsymbol{R}_s=P_1\boldsymbol{e}_1\boldsymbol{e}_1^{\mathrm{H}}$$

Because $\boldsymbol{R}_s$ has a rank of one, then the matrix $\boldsymbol{R}_s$ has only one non-zero eigenvalue. And we also have,

$$\boldsymbol{R}_s\boldsymbol{e}_1=P_1(\boldsymbol{e}_1\boldsymbol{e}_1^{\mathrm{H}})\boldsymbol{e}_1=P_1\boldsymbol{e}_1(\boldsymbol{e}_1^{\mathrm{H}}\boldsymbol{e}_1)=MP_1\boldsymbol{e}_1$$

This means that the non-zero eigenvalue is MP_1, and $\boldsymbol{e}_1$ is the corresponding eigen-vector. Besides, since $\boldsymbol{R}_s$ is Hermitian, then the remaining eigenvectors, $\boldsymbol{v}_2,\boldsymbol{v}_3,\cdots,\boldsymbol{v}_M$ are orthogonal to $\boldsymbol{e}_1$, namely

$$\boldsymbol{e}_1^{\mathrm{H}}\boldsymbol{v}_i=0;i=2,3,\cdots,M \tag{4.80}$$

Furthermore, if we define λ_i^s as the eigenvalues of $\boldsymbol{R}_s$, then we have

$$\boldsymbol{R}_x\boldsymbol{v}_i=(\boldsymbol{R}_s+\sigma_w^2\boldsymbol{I})\boldsymbol{v}_i=\lambda_i^s\boldsymbol{v}_i+\sigma_w^2\boldsymbol{v}_i=(\lambda_i^s+\sigma_w^2)\boldsymbol{v}_i \tag{4.81}$$

Then, the eigen-vectors of $\boldsymbol{R}_x$ are same as those of $\boldsymbol{R}_s$:

$$\lambda_i=\lambda_i^s+\sigma_w^2$$

Thus, the largest eigenvalue of $\boldsymbol{R}_x$ is

$$\lambda_{\max}=MP_1+\sigma_w^2$$

And the remaining $M-1$ eigenvalues are equal to σ_w^2.

Finally, we can extract all of the parameters of interest about $x(n)$ from the eigenvalues and eigen-vectors of $\boldsymbol{R}_x$:

1. Perform an eigen-decomposition of the autocorrelation matrix $\boldsymbol{R}_x$. The largest eigenvalue $\lambda_{\max}=MP_1+\sigma_w^2$ and the remaining eigenvalues will be equal to σ_w^2.

2. Use the eigenvalues of $\boldsymbol{R}_x$ to solve for the power $P_1=|A_1|^2$ and the noise variance as follows:

$$P_1=\frac{1}{M}(\lambda_{\max}-\sigma_w^2)$$

$$\sigma_w^2=\lambda_{\min}$$

3. Determine the frequency ω_1 from the eigenvector $\boldsymbol{v}_{\max}$. The components of the eigen-vectors are numbered from $v_i(0)$ to $v_i(M-1)$, then ω_1 is associated with the largest eigenvalue, and using the second coefficient of $\boldsymbol{v}_{\max}$ we can obtain:

$$\omega_i=\arg\{v_{\max}(1)\}$$

where $\arg\{\cdot\}$ means the argument of a complex number.

Once the P_1, σ_w^2 and ω_1 are determined, the $r(k)$ and its Fourier transform, namely the power spectral density of $x(n)$, can be achieved.

Instead of estimating the frequency of the complex exponential from a single eigenvector, we can also use a weighted average as follows. Let $\boldsymbol{v}_i$ be an eigen-vector of $\boldsymbol{R}_x$ with eigenvalue σ_w^2, and let $v_i(k)$ be the k th component of $\boldsymbol{v}_i$. Then the DTFT of the coefficients in $\boldsymbol{v}_i$ is,

$$V_i(\mathrm{e}^{\mathrm{j}\omega})=\sum_{k=0}^{M-1}v_i(k)\mathrm{e}^{-\mathrm{j}k\omega}=\boldsymbol{e}^{\mathrm{H}}\boldsymbol{v}_i \tag{4.82}$$

From the equation (4.80), we can know that $V_i(\mathrm{e}^{\mathrm{j}\omega})$ equals zero at $\omega=\omega_1$. Thus, if we define a *frequency estimation function* as follows:

$$\hat{P}_i(e^{j\omega}) = \frac{1}{\left|\sum_{k=0}^{M-1} v_i(k)e^{-jk\omega}\right|^2} = \frac{1}{|\boldsymbol{e}^H \boldsymbol{v}_i|^2} \tag{4.83}$$

Then $\hat{P}_i(e^{j\omega})$ can be used to estimate ω_1 by the location of its peak. Furthermore, we can consider using a weighted average of all the eigen-vectors instead of using only one single eigen-vector,

$$\hat{P}(e^{j\omega}) = \frac{1}{\sum_{i=2}^{M} \alpha_i |\boldsymbol{e}^H \boldsymbol{v}_i|^2} \tag{4.84}$$

where α_i are some appropriately chosen constants.

Now, let us consider the general case of a WSS process consisting of p different complex exponentials in white noise. The autocorrelation sequence is

$$r(k) = \sum_{i=1}^{p} P_i e^{jk\omega_i} + \sigma_w^2 \delta(k)$$

Therefore, the autocorrelation matrix is

$$\boldsymbol{R}_x = \boldsymbol{R}_s + \boldsymbol{R}_n = \sum_{i=1}^{p} P_i \boldsymbol{e}_i \boldsymbol{e}_i^H + \sigma_w^2 \boldsymbol{I} \tag{4.85}$$

where we can get a set of p linearly independent vectors,

$$\boldsymbol{e}_i = [1, e^{j\omega_i}, e^{j2\omega_i}, \cdots, e^{j(M-1)\omega_i}]^T; i = 1,2,\cdots,p$$

Equation (4.85) can be written as

$$\boldsymbol{R}_x = \boldsymbol{EPE}^H + \sigma_w^2 \boldsymbol{I} \tag{4.86}$$

where $\boldsymbol{E} = [\boldsymbol{e}_1, \cdots, \boldsymbol{e}_p]$, and $\boldsymbol{P} = diag\{P_1, \cdots, P_p\}$. The eigenvalues and eigen-vectors of $\boldsymbol{R}_x$ may again be divided into two groups: the signal eigen-vectors $\boldsymbol{v}_1, \cdots, \boldsymbol{v}_p$ that have eigenvalues ($\lambda_i = \lambda_i^s + \sigma_w^2$, where λ_i^s are the eigenvalues of $\boldsymbol{R}_s$ with rank p) greater than σ_w^2, and the noise eigen-vectors $\boldsymbol{v}_{p+1}, \cdots, \boldsymbol{v}_M$ that have eigenvalues equal to σ_w^2. We may use the spectral theorem to decompose $\boldsymbol{R}_x$:

$$\boldsymbol{R}_x = \sum_{i=1}^{p} (\lambda_i^s + \sigma_w^2) \boldsymbol{v}_i \boldsymbol{v}_i^H + \sum_{i=p+1}^{M} \sigma_w^2 \boldsymbol{v}_i \boldsymbol{v}_i^H$$

This decomposition can be written as

$$\boldsymbol{R}_x = \boldsymbol{V}_{SS} \boldsymbol{V}_S^H + \boldsymbol{V}_{nn} \boldsymbol{V}_n^H$$

where $\boldsymbol{V}_S = [\boldsymbol{v}_1, \boldsymbol{v}_2, \cdots, \boldsymbol{v}_p]$, $\boldsymbol{V}_n = [\boldsymbol{v}_{p+1}, \boldsymbol{v}_{p+2}, \cdots, \boldsymbol{v}_M]$, and the double subscript means that the matrix contains the eigenvalues $\lambda_i = \lambda_i^s + \sigma_w^2$ and $\lambda_i = \sigma_w^2$. Thus, like equation (4.84), the frequencies can be estimated using a frequency estimation function as follows:

$$\hat{P}(e^{j\omega}) = \frac{1}{\sum_{i=p+1}^{M} \alpha_i |\boldsymbol{e}^H \boldsymbol{v}_i|^2} \tag{4.87}$$

In the following sections, we will develop several different kinds of frequency estimation algorithms based on equation (4.87). We first introduce the Pisarenko harmonic decomposition, which uses a frequency estimator of this form with $M = p + 1$ and $\alpha_M = 1$.

Pisarenko harmonic decomposition. As for the Pisarenko harmonic decomposition, it is assumed that $x(n)$ is a sum of p complex exponentials in white noise, and p is known. It is assumed that $p+1$ values of the autocorrelation sequence are either known or have been estimated, with a $(p+1) \times (p+1)$ autocorrelation matrix. Denoting the noise eigen-vector by $\boldsymbol{v}_{\min}$, it follows that $\boldsymbol{v}_{\min}$ will be orthogonal to each of the signal vectors $\boldsymbol{e}_i$ (recall the equation (4.80)),

$$\boldsymbol{e}_i^{\mathrm{H}} \boldsymbol{v}_{\min} = \sum_{k=0}^{p} v_{\min}(k) \mathrm{e}^{-\mathrm{j}k\omega_1} = 0; i = 1, 2, \cdots, p \tag{4.88}$$

Therefore,

$$V_{\min}(\mathrm{e}^{\mathrm{j}\omega}) = \sum_{k=0}^{p} v_{\min}(k) \mathrm{e}^{-\mathrm{j}k\omega}$$

is equal to zero at each of the complex exponential frequencies ω_i for $i = 1, 2, \cdots, p$. Then, the z-transform of the noise eigen-vector referred to as an eigen-filter, has p zeros on the unit circle,

$$V_{\min}(z) = \sum_{k=0}^{p} v_{\min}(k) z^{-k} = \prod_{k=1}^{p} (1 - \mathrm{e}^{\mathrm{j}\omega_k} z^{-1}) \tag{4.89}$$

we can also form the frequency estimation

$$\hat{P}_{\mathrm{PHD}}(\mathrm{e}^{\mathrm{j}\omega}) = \frac{1}{|\boldsymbol{e}^{\mathrm{H}} \boldsymbol{v}_{\min}|^2} \tag{4.90}$$

which is a special case of equation (4.87) with $M = p + 1$ and $\alpha_M = 1$. Since $\hat{P}_{\mathrm{PHD}}(\mathrm{e}^{\mathrm{j}\omega})$ will be large at the frequencies of the complex exponentials, the locations of the peaks in $\hat{P}_{\mathrm{PHD}}(\mathrm{e}^{\mathrm{j}\omega})$ can be used as frequency estimates. Although written in the form of a power spectrum, $\hat{P}_{\mathrm{PHD}}(\mathrm{e}^{\mathrm{j}\omega})$ is called a pseudo-spectrum (or an eigen-spectrum), since it does not contain any information about the power in the complex exponentials.

MUSIC. Multiple Signal Classification method (MUSIC) technique for frequency estimation is an improvement to the Pisarenko harmonic decomposition. Assume $x(n)$ to be a random process consisting of p complex exponentials in white noise with a variance of σ_w^2, and $\boldsymbol{R}_x$ to be a $M \times M$ autocorrelation matrix of $x(n)$ with $M > p + 1$ (If $M = p + 1$, then the MUSIC algorithm is equivalent to Pisarenko's method). If the eigenvalues of $\boldsymbol{R}_x$ and the corresponding eigen-vectors are arranged as $\lambda_1 \geqslant \lambda_2 \geqslant \cdots \geqslant \lambda_M$ and $\boldsymbol{v}_1, \boldsymbol{v}_2, \cdots, \boldsymbol{v}_M$, we can divide the eigen-vectors into two groups: the p signal eigen-vectors corresponding to the p largest eigenvalues, and the $M - p$ noise eigen-vectors with eigenvalues equal to σ_w^2. Therefore, the eigen-vectors of $\boldsymbol{R}_x$ are of length M, and for each noise eigen-vector we have:

$$V_i(z) = \sum_{k=0}^{M-1} v_i(k) z^{-k}; i = p + 1, \cdots, M$$

Furthermore, p of these roots lie on the unit circle at the frequencies of the complex exponentials, and the eigen-spectrum is as follows:

$$|V_i(\mathrm{e}^{\mathrm{j}\omega})|^2 = \frac{1}{\left| \sum_{k=0}^{M-1} v_i(k) \mathrm{e}^{-\mathrm{j}k\omega} \right|^2}$$

And the frequency estimation function is

$$\hat{P}_{\mathrm{MU}}(\mathrm{e}^{\mathrm{j}\omega}) = \frac{1}{\sum_{i=p+1}^{M} |\boldsymbol{e}^{\mathrm{H}} \boldsymbol{v}_i|^2} \tag{4.91}$$

Then, the frequencies of the complex exponentials are then taken as the locations of the p largest peaks in $\hat{P}_{\mathrm{MU}}(\mathrm{e}^{\mathrm{j}\omega})$.

An alternative of MUSIC is to use a method named as root MUSIC, which involves rooting a polynomial instead of searching for the peaks of $\hat{P}_{\mathrm{MU}}(\mathrm{e}^{\mathrm{j}\omega})$.

The z -transform equivalent of equation (4.91) is

$$\hat{P}_{\mathrm{MU}}(z) = \frac{1}{\sum_{i=p+1}^{M} V_i(z) V_i^*(1/z^*)}$$

Then we can take the frequency estimates to be the angles of the p roots of the polynomial

$$D(z) = \sum_{i=p+1}^{M} V_i(z) V_i^*(1/z^*) \tag{4.92}$$

Other eigenvector methods. There are many other eigenvector methods proposed for estimating the frequencies of the complex exponentials in noise. In addition to Pisarenko and MUSIC methods, we introduce the EigenVector (EV) method.

The EigenVector (EV) method is close to the MUSIC method, which estimates the exponential frequencies from the peaks of the eigen-spectrum:

$$\hat{P}_{\mathrm{EV}}(\mathrm{e}^{\mathrm{j}\omega}) = \frac{1}{\sum_{i=p+1}^{M} \frac{1}{\lambda_i} |\boldsymbol{e}^{\mathrm{H}} \boldsymbol{v}_i|^2} \tag{4.93}$$

where λ_i is the eigenvalue corresponding to the eigen-vector $\boldsymbol{v}_i$. In addition, the Eigen-Vector method differs from MUSIC algorithm with fewer spurious peaks.

4.4.4 Principal Components Spectrum Estimation

In the section before, we found that the orthogonality of the signal and noise subspaces could be used to estimate the frequencies of p complex exponentials in white noise. Since these methods mainly use the vectors in the noise subspace, they are usually called as noise subspace methods.

In this section, we consider the methods mainly using the vectors in the signal subspace, which are based on a principal components analysis of the autocorrelation matrix, and usually called as signal subspace methods.

Let $\boldsymbol{R}_x$ be an $M \times M$ autocorrelation matrix of a process consisted of p complex exponentials in white noise. With an eigen-decomposition of $\boldsymbol{R}_x$, we have

$$\boldsymbol{R}_x = \sum_{i=1}^{M} \lambda_i \boldsymbol{v}_i \boldsymbol{v}_i^{\mathrm{H}} = \sum_{i=1}^{p} \lambda_i \boldsymbol{v}_i \boldsymbol{v}_i^{\mathrm{H}} + \sum_{i=p+1}^{M} \lambda_i \boldsymbol{v}_i \boldsymbol{v}_i^{\mathrm{H}} \tag{4.94}$$

where the eigenvalues have been arranged in decreasing order $\lambda_1 \geqslant \lambda_2 \geqslant \cdots \geqslant \lambda_M$. Since the second term in equation (4.94) is due only to the noise, we may form a reduced rank approximation to the signal autocorrelation matrix $\boldsymbol{R}_s$ by retaining only the principal eigenvectors of $\boldsymbol{R}_x$,

$$\hat{\boldsymbol{R}}_s = \sum_{i=1}^{p} \lambda_i \boldsymbol{v}_i \boldsymbol{v}_i^{\mathrm{H}} \tag{4.95}$$

The principal components approximation may then be used in the place of $\boldsymbol{R}_x$ in a spectral estimator, such as the minimum variance method or the maximum entropy method.

This approach is to filter out a portion of the noise, and enhance the estimate of the spectral component relating to the signal alone. Another way to view this approach is to place a constraint on the autocorrelation matrix. Considering a process consisted of p complex exponentials in noise, since the rank of the autocorrelation matrix relating to $\boldsymbol{R}_s$ is p, then the principal components can simply impose the p -rank constraint on $\boldsymbol{R}_x$.

In the following subsections, we discuss how a principal components analysis of the autocorrelation matrix may be used to form a principal components spectrum estimate.

Blackman-Tukey frequency estimation. We can perform the Blackman-Tukey estimate of the power spectrum by taking the discrete-time Fourier transform of a windowed autocorrelation sequence as follows

$$\hat{P}_{\mathrm{BT}}(\mathrm{e}^{\mathrm{j}\omega}) = \sum_{k=-M}^{M} \hat{r}_x(k) w(k) \mathrm{e}^{-\mathrm{j}k\omega}$$

If $w(k)$ is a Bartlett window, then the Blackman-Tukey estimate can be written in terms of the autocorrelation matrix $\boldsymbol{R}_x$ as follows

$$\hat{P}_{\mathrm{BT}}(\mathrm{e}^{\mathrm{j}\omega}) = \frac{1}{M}\sum_{k=-M}^{M} (M - |k|)\hat{r}_x(k) \mathrm{e}^{-\mathrm{j}k\omega} = \frac{1}{M}\boldsymbol{e}^{\mathrm{H}} \boldsymbol{R}_x \boldsymbol{e} \tag{4.96}$$

With an eigen-decomposition of the autocorrelation matrix, $\hat{P}_{\mathrm{BT}}(\mathrm{e}^{\mathrm{j}\omega})$ can be written as

$$\hat{P}_{\mathrm{BT}}(\mathrm{e}^{\mathrm{j}\omega}) = \frac{1}{M}\sum_{i=1}^{M} \lambda_i |\boldsymbol{e}^{\mathrm{H}} \boldsymbol{v}_i|^2$$

If we know that $x(n)$ is consisted of p complex exponentials in white noise, and the eigenvalues of $\boldsymbol{R}_x$ are arranged in decreasing order, then a principal components version of the spectrum estimate is as follows

$$\hat{P}_{\mathrm{PC-BT}}(\mathrm{e}^{\mathrm{j}\omega}) = \frac{1}{M}\boldsymbol{e}^{\mathrm{H}} \hat{\boldsymbol{R}}_s \boldsymbol{e} = \frac{1}{M}\sum_{i=1}^{p} \lambda_i |\boldsymbol{e}^{\mathrm{H}} \boldsymbol{v}_i|^2 \tag{4.97}$$

Minimum variance frequency estimation. Considering the autocorrelation sequence $r_x(k)$ of a process $x(n)$ for lags $|k| \leqslant M$, the M -order minimum variance spectrum estimate is as follows

$$\hat{P}_{\mathrm{MV}}(\mathrm{e}^{\mathrm{j}\omega}) = \frac{M}{\boldsymbol{e}^{\mathrm{H}} \boldsymbol{R}_x^{-1} \boldsymbol{e}} \tag{4.98}$$

With an eigen-decomposition of the autocorrelation matrix, the inverse of $\boldsymbol{R}_x$ is

$$\boldsymbol{R}_x^{-1} = \sum_{i=1}^{p} \frac{1}{\lambda_i} \boldsymbol{v}_i \boldsymbol{v}_i^{\mathrm{H}} + \sum_{i=p+1}^{M} \frac{1}{\lambda_i} \boldsymbol{v}_i \boldsymbol{v}_i^{\mathrm{H}} \tag{4.99}$$

where p is the number of complex exponentials. Retaining only the first p principal components of $\boldsymbol{R}_x^{-1}$, we can obtain the principal components minimum variance estimate

$$\hat{P}_{\mathrm{PC-MV}}(\mathrm{e}^{\mathrm{j}\omega}) = \frac{M}{\sum_{i=1}^{p} \frac{1}{\lambda_i} |\boldsymbol{e}^{\mathrm{H}} \boldsymbol{v}_i|^2} \tag{4.100}$$

Autoregressive frequency estimation. Autoregressive spectrum estimation involves finding the solution to a set of linear equations as follows

$$\boldsymbol{R}_x \boldsymbol{a}_M = \varepsilon_M \boldsymbol{u}_1 \tag{4.101}$$

where $\boldsymbol{R}_x$ is an $(M+1)\times(M+1)$ autocorrelation matrix. From the solution to these equations

$$\boldsymbol{a}_M = \varepsilon_M \boldsymbol{R}_x^{-1} \boldsymbol{u}_1$$

we can get an estimate of the spectrum as follows

$$\hat{P}_{\mathrm{AR}}(\mathrm{e}^{\mathrm{j}\omega}) = \frac{|b(0)|^2}{|\boldsymbol{e}^{\mathrm{H}} \boldsymbol{a}_M|^2}$$

where $b(0)$ is chosen to be a constant so that $|b(0)|^2 = \varepsilon_M$. If $x(n)$ is known to consist of p complex exponentials in noise, then we may form a principal components solution to equation (4.101) as follows

$$\boldsymbol{a}_{pc} = \varepsilon_M \left(\sum_{i=1}^{p} \frac{1}{\lambda_i} \boldsymbol{v}_i \boldsymbol{v}_i^{\mathrm{H}} \right) \boldsymbol{u}_1$$

Furthermore, we also have

$$\boldsymbol{a}_{pc} = \varepsilon_M \sum_{i=1}^{p} \frac{v_i^*(0)}{\lambda_i} \boldsymbol{v}_i = \varepsilon_M \sum_{i=1}^{p} \alpha_i \boldsymbol{v}_i$$

where $v_i(0)$ is the first element of the normalized eigenvector $\boldsymbol{v}_i$ and $\alpha_i = v_i^*(0)/\lambda_i$. Thus, if we set $|b(0)|^2 = \varepsilon_M$, then the principal components autoregressive spectrum estimate is as follows

$$\hat{P}_{\mathrm{PC\text{-}AR}}(\mathrm{e}^{\mathrm{j}\omega}) = \frac{1}{\left| \sum_{i=1}^{p} \alpha_i \boldsymbol{e}^{\mathrm{H}} \boldsymbol{v}_i \right|^2} \tag{4.102}$$

4.5 Summary

We discussed the nonparametric methods based on computing the discrete-time Fourier transform of an estimate of the autocorrelation sequence. The first method was the periodogram, which is evaluated from the DFT of the given process values. However, the periodogram is not a consistent estimate of the power spectrum. Thus, we researched some modifications of the periodogram to improve the statistical properties. These modifications included applying a window to the data, periodogram averaging, and periodogram smoothing. Although periodogram averaging and periodogram smoothing can provide a consistent estimate of the power spectrum, they generally do not suit well for short data records, and are limited in resolving closely spaced narrowband processes with the number of data samples limited. On the other hand, the advantage of these methods is that they do not make any assumptions or place any constraints on the process and can be applied on any type of process.

Next, we discussed the parametric methods of spectrum estimation. With a parametric approach, the first step is to select an appropriate model for the process. This selection may be based on a priori knowledge about how the process is generated or on experimental results indicating that a particular model "works well". Once a model has been selected, the next step is to estimate the model parameters from the given data. For example, if $x(n)$ is assumed to be an autoregressive process, then the covariance methods or some other algorithms may be used to estimate the all-pole parameters. The final step is to estimate the power spectrum by incorporating the estimated parameters into the parametric form for the spectrum.

Although it is possible to significantly improve the performance of the spectrum estimate with a parametric approach, it is important that the model that is used should be consistent with the process that is being analyzed. Otherwise, inaccurate or misleading spectrum estimates may result.

After the discussion of the parametric methods, we introduced some other methods of the spectrum estimation. We derived the minimum variance method, which can be considered as a data-adaptive modification to the periodogram. The main idea of this approach is to design a filter bank of bandpass filters first, and then measure the power in the processes produced at the output of each filter, and finally estimate the power spectrum by dividing this power estimate by the bandwidth of the filter. Note that the minimum variance spectrum estimate provides higher resolution than the periodogram and Blackman-Tukey methods.

Then the maximum entropy method was derived, which estimates the spectrum using a maximum entropy extrapolation of a given partial autocorrelation sequence. In other words, given $r_x(k)$ for $|k| \leqslant p$, the values of $r_x(k)$ for $|k| > p$ are found that make the underlying process as white or as random as possible. What we discovered, however, is that this is equivalent to finding an all-pole model for the process that is consistent with the given autocorrelation sequence, and then computing the power spectrum from the all-pole model.

Finally, the set of discussed techniques are those that assume a harmonic model for the process, namely that $x(n)$ is a sum of complex exponentials or sinusoids in white noise. The goal of these processes is to estimate the frequencies of the complex exponentials and determine the powers next.

Two different kinds of approaches to the frequency estimation were considered here. The first defines a frequency estimation function to produce peaks at the frequencies of the complex exponentials, and these frequency estimation functions are designed to take advantage of the fact that the signal and noise subspaces are orthogonal. The second kind of methods use a principal components analysis of the autocorrelation matrix, and the resulting reduced-rank approximation of the autocorrelation matrix is used in a spectrum estimation method such as the minimum variance method or the maximum entropy method.

Many different approaches of spectrum estimation have been presented in this chapter, but there are many other methods that have been proposed and are superior to the methods described here under certain conditions or set of assumptions.

Exercises

1. Bartlett's method is a kind of nonparametric methods to estimate the power spectrum of a process from a sequence of $N = 2{,}000$ samples.

(1) What is the minimal length L that could be used for each sequence if we want to have a

resolution of $\Delta f = 0.005$?

(2) Explain why it will not be advantageous to increase L beyond the value found in (a).

(3) The quality factor of a spectrum estimate is defined to be the inverse of the variability as follows

$$Q = 1/V$$

Using Bartlett's method, what is the minimal number of data samples, N, that are necessary to achieve a resolution of $\Delta f = 0.005$, and a quality factor that is 5 times larger than that of the periodogram?

Solution:

(1) Since $\Delta f = 0.9/L$ then

$$L = \frac{0.89}{\Delta f} = \frac{0.9}{0.005} = 180$$

(2) Increasing L will increase the resolution, but it will also result in a decrease in the number of segments that may be averaged. This, in turn, will increase the variance of the spectrum estimate.

(3) For the periodogram, the quality factor is $Q_{\text{per}} = 1/V_{\text{per}} = 1$. For the Bartlett's method, the quality factor is $Q_B = 1/V_B = K$. Thus, if we want $Q_{\text{per}}/Q_B \geqslant 5$, then we must have $K \geqslant 5$. With $M =$ 180 for $\Delta f = 0.005$, then we must have

$$N = KM \geqslant 5 \times 180 = 900$$

The Periodogram:

```
function Px=periodgram(x, n1, n2)
%
    x=x(:);
    if nargin==1
       n1=1; n2=length(x); end;
    Px=abs(fft(x(n1:n2), 1024)).^2/(n2-n1+1);
    Px(1)=Px(2);
```

Bartlett's Method: Periodogram Averaging:

```
    function Px=bart(x, nsect)
%
    L=floor(length(x)/nsect);
    Px=0;
    n1=1;
    for i=1:nsect
       Px=Px+periodgram(x(n1:n1+L-1))/nsect;
       N1=n1+L;
```

2. The estimated autocorrelation sequence of a random process $x(n)$ for lags $k=0,1,2,3,4$ are

$$r_x(0)=2;\quad r_x(1)=1;\quad r_x(2)=1;\quad r_x(3)=0.5;\quad r_x(4)=0;$$

Estimate the power spectrum of $x(n)$ for each of the following conditions.

(1) $x(n)$ is an AR(2) process.

(2) $x(n)$ is an MA(2) process.

(3) $x(n)$ is an ARMA(1,1) process.

Solution:

(1) For an AR(2) process, we want to find a second-order AR model. This is done by solving the normal equations

$$\begin{bmatrix} r_x(0) & r_x(1) \\ r_x(1) & r_x(0) \end{bmatrix}\begin{bmatrix} a(1) \\ a(2) \end{bmatrix} = -\begin{bmatrix} r_x(1) \\ r_x(2) \end{bmatrix}$$

For the given autocorrelation sequence, these become

$$\begin{bmatrix} 2 & 1 \\ 1 & 2 \end{bmatrix}\begin{bmatrix} a(1) \\ a(2) \end{bmatrix} = -\begin{bmatrix} 1 \\ 1 \end{bmatrix}$$

Thus, the coefficients are

$$\begin{bmatrix} a(1) \\ a(2) \end{bmatrix} = -\frac{1}{3}\begin{bmatrix} 1 \\ 1 \end{bmatrix}$$

With a modeling error

$$\varepsilon_2 = r_x(0) + a(1)r_x(1) + a(2)r_x(2) = \frac{4}{3}$$

Thus

$$A(e^{j\omega}) = 1 - \frac{1}{3}e^{-j\omega} - - \frac{1}{3}e^{-2j\omega}$$

The power spectrum is

$$\hat{P}_x(e^{j\omega}) = \frac{\frac{4}{3}}{\left|1 - \frac{1}{3}e^{-j\omega} - \frac{1}{3}e^{-2j\omega}\right|^2}$$

(2) For an MA(2) process

$$\hat{P}_{\text{MA}}(e^{j\omega}) = \sum_{k=-2}^{2} r_x(k)e^{-jk\omega} = 2 + 2\cos\omega + 2\cos 2\omega$$

(3) For an ARMA(1,1) process, we must solve the Yule-Walker equations

$$\begin{bmatrix} r_x(0) & r_x(1) \\ r_x(1) & r_x(0) \\ r_x(2) & r_x(1) \end{bmatrix}\begin{bmatrix} 1 \\ a(1) \end{bmatrix} = \begin{bmatrix} c(0) \\ c(1) \\ 0 \end{bmatrix}$$

The coefficient $a(1)$ is found from the last equation as follows,

$$a(1) = -r_x(2)/r_x(1) = -1$$

Solving for $c(0)$ and $c(1)$ we can obtain

$$\begin{bmatrix} c(0) \\ c(1) \end{bmatrix} = \begin{bmatrix} r_x(0) & r_x(1) \\ r_x(1) & r_x(0) \end{bmatrix} \begin{bmatrix} 1 \\ a(1) \end{bmatrix} = \begin{bmatrix} 2 & 1 \\ 1 & 2 \end{bmatrix} \begin{bmatrix} 1 \\ -1 \end{bmatrix} = \begin{bmatrix} 1 \\ -1 \end{bmatrix}$$

Then, we have

$$B(z)B(z^{-1}) = A(z^{-1})C(z) = [1 + a(1)z] \sum_{k=-\infty}^{\infty} c(k)z^{-k}$$
$$= \cdots + [c(-1) - c(0)]z + [c(0) - c(1)] + [c(1) - c(2)]z^{-1} + \cdots$$

Since $B(z)B(z^{-1})$ is symmetric and order one in z and z^{-1}, with $c(0) = 1$ and $c(1) = -1$, it follows that

$$B(z)B(z^{-1}) = -z + 2 - z^{-1}$$

Thus, the power spectrum estimate is

$$\hat{P}_x(e^{j\omega}) = \frac{|B(e^{j\omega})|^2}{|A(e^{j\omega})|^2} = \frac{2 - 2\cos\omega}{2 - 2\cos\omega} = 1$$

References

[1] S. Kay, Modern Spectrum Estimation: Theory and Applications, Prentice-Hall, Englewood Cliffs, NJ, 1988.

[2] S. Kay, "Noise compensation for autoregressive spectral estimates," IEEE Trans. Acoust., Speech, Sig. Proc., vol. ASSP-28, pp. 292-303, June 1980.

[3] S. Kay and S. L. Marple Jr., "Sources and remedies for spectral line splitting in autoregressive spectrum analysis," Proc. 1979 Int. Conf. on Acoust., Speech, Sig. Proc, pp. 151-154, 1979.

[4] R. H. Jones, "Identification and autoregressive spectrum estimation," IEEE Trans. Autom. Control, vol. AC-19, pp. 894-897, Dec. 1974.

[5] D. M. Thomas and M. H. Hayes. "A novel data-adaptive power spectrum estimation technique," Proc. 1987 Int. Conf. on Acoustics, speech sig. Proc., pp. 1589-1595, April. 1987.

Chapter 5

Optimum Filters

5.1 Introduction

Estimating one signal from another is one of the most critical problems in signal processing, and it embraces a wide range of practical applications. Norbert Wiener pioneered the design of this type of filter problem, so now the filter designed based on the Minimum Mean Square Error criterion (MMSE) is called a Wiener filter, and its basic principle is to take the Mean Square Error (MSE) as the cost function, design the filter that minimizes the cost function, and recover the original signal $d(n)$ from the noisy observation signal $x(n)$

$$x(n) = d(n) + v(n)$$

Assuming that both $d(n)$ and $v(n)$ are wide-sense stationary random processes, Wiener considers the design of filters that can produce a Minimum Mean Square Error estimate of $d(n)$, labeled $\hat{d}(n)$. And when defining

$$\xi = E\{ |e(n)|^2 \}$$

where

$$e(n) = d(n) - \hat{d}(n)$$

the filters' purpose is to minimize ξ. We start this chapter by considering the general problem of Wiener filtering, i. e., designing a linear shift-invariant filter $W(z)$ that, after filtering a given signal $x(n)$, generates the minimum mean square estimate $\hat{d}(n)$ of desired signal $d(n)$.

Depending on the definition of the relationship between the signals $x(n)$ and $d(n)$, several different significant problems can be incorporated into the Wiener filtering framework and some of the issues that will be considered in this chapter include:

1. Filtering. It is a classical problem considered by Wiener in which we are given $x(n) = d(n) + v(n)$ and wish to estimate $d(n)$ using a causal filter, i. e., from the current and past values of $x(n)$.

2. Smoothing. Same as the filtering problem, except that the filter is allowed to be non-causal under this application. For example, a Wiener smoothing filter can be designed to estimate $d(n)$ from $x(n) = d(n) + v(n)$ using all available data.

3. Prediction. If the desired $d(n) = x(n+1)$ and $W(z)$ is a causal filter, then the Wiener filter becomes a linear predictor. In this case, the filter is to produce a prediction of $x(n+1)$ (also approximated as an estimate) based on a linear combination of previous values of $x(n)$.

4. Deconvolution. When $x(n) = d(n) * g(n) + v(n)$ with $g(n)$ being the unit impulse response of a linear shift-invariant filter, the Wiener filter becomes a deconvolution filter.

The rest of this chapter organizes as follows, Section 5.2 presents the design of FIR Wiener filter, and the main result here will be the derivation of the discrete form of the Wiener-Hopf equations, which specifies the coefficients of the optimum filter (by minimizing MSE). Then, we derive the solutions of the Wiener-Hopf equations for the cases of filtering, smoothing, prediction, and noise cancellation respectively. The design of IIR Wiener filters is presented in Section 5.3. First, in Section 5.3.1 we solve the noncausal Wiener filtering problem. Then, in Section 5.3.2, we describe how to design a causal Wiener filter.

In addition, we arrange a part of the derivation to explore how to use the observations to filter out the noise as much as possible when the first or second order statistical properties of the signal such as the expectation or the variances are not available. Context involves the Least Square Error criterion (LSE) and is placed in Section 5.4. We will present the FIR filter design problem under the Least Square (LS) estimation principle as well as the solution of the LSE norm equation and the computation of the estimates using the SVD method. The final part, Section 5.5 introduces recursive approaches to signal estimation and derives the discrete Kalman filter.

5.2 The FIR Wiener Filter

In this section we introduce the design of the FIR Wiener filter to produce a Minimum Mean Square estimate for a given process $d(n)$ by filtering a set of observations of a statistically related process $x(n)$. First assume that $x(n)$ and $d(n)$ are jointly wide-sense stationary with known autocorrelations $\boldsymbol{R}_x(k)$ and $\boldsymbol{R}_d(k)$, and known cross-correlation $\boldsymbol{r}_{dx}(k)$. Next the unit impulse response of the Wiener filter is denoted by $w(n)$ and assumed to be a $(p-1)$-st order filter with a system function of

$$W(z) = \sum_{n=0}^{p-1} w(n) z^{-n}$$

The input of the filter is $x(n)$, and the desired output denoted by $\hat{d}(n)$, is the convolution of $w(n)$ with $x(n)$,

$$\hat{d}(n) = \sum_{l=0}^{p-1} w(l) x(n-l) \tag{5.1}$$

The Wiener filter is designed to find the filter coefficients $w(k)$ that minimize the Mean Square Error

$$\xi = E\{|e(n)|^2\} = E\{|d(n) - \hat{d}(n)|^2\} \tag{5.2}$$

and to find the set of filter coefficients to minimize ξ. It is necessary and reasonable to let the derivative of ξ, with concerning to $w^*(k), k = 0, 1, \cdots, p-1$ be equal to zero

$$\frac{\partial \xi}{\partial w^*(k)} = \frac{\partial E\{e(n)e^*(n)\}}{\partial w^*(k)} = E\left\{e(n)\frac{\partial e^*(n)}{\partial w^*(k)}\right\} = 0 \tag{5.3}$$

With

$$e(n) = d(n) - \sum_{l=0}^{p-1} w(l)x(n-l) \tag{5.4}$$

$$\frac{\partial e^*(n)}{\partial w^*(k)} = -x^*(n-k)$$

Equation (5.3) becomes

$$E\{e(n)x^*(n-k)\} = 0, k = 0,1,\cdots,p-1 \tag{5.5}$$

which is known as the orthogonality principle or the projection theorem. Then substituting equation (5.4) into equation (5.5) yields the form of the equation

$$E\{d(n)x^*(n-k)\} - \sum_{l=0}^{p-1} w(l)E\{x(n-l)x^*(n-k)\} = 0 \tag{5.6}$$

Finally, since it is assumed that $x(n)$ and $d(n)$ are jointly wide-sense stationary processes, under the relation with $r_x(k-l) = E\{x(n-l)x^*(n-k)\}$ and $r_{dx}(k) = E\{d(n)x^*(n-k)\}$, equation (5.6) becomes

$$\sum_{l=0}^{p-1} w(l)r_x(k-l) = r_{dx}(k), k = 0,1,\cdots,p-1 \tag{5.7}$$

which is a set of p linear equations in the p unknowns $w(k), k = 0,1,\cdots,p-1$. Convert equation (5.7) to the form of the matrix as

$$\begin{bmatrix} r_x(0) & r_x^*(1) & r_x^*(2) & \cdots & r_x^*(p-1) \\ r_x(1) & r_x(0) & r_x^*(1) & \cdots & r_x^*(p-2) \\ r_x(2) & r_x(1) & r_x(0) & \cdots & r_x^*(p-3) \\ \vdots & \vdots & \vdots & \ddots & \vdots \\ r_x(p-1) & r_x(p-2) & r_x(p-3) & \cdots & r_x(0) \end{bmatrix} \begin{bmatrix} \omega(0) \\ \omega(1) \\ \omega(2) \\ \vdots \\ \omega(p-1) \end{bmatrix} = \begin{bmatrix} r_{dx}(0) \\ r_{dx}(1) \\ r_{dx}(2) \\ \vdots \\ r_{dx}(p-1) \end{bmatrix} \tag{5.8}$$

At this point, we derive the matrix form of the Wiener-Hopf equations equation (5.8), and it could be written more concisely as

$$\boldsymbol{R}_x \boldsymbol{w} = \boldsymbol{r}_{dx} \tag{5.9}$$

where $\boldsymbol{R}_x$ is a $p \times p$ Hermitian Toeplitz matrix of autocorrelations, $\boldsymbol{w}$ is the vector of filter coefficients, and $\boldsymbol{r}_{dx}$ is the vector of cross-correlations between the desired signal $d(n)$ and the observed signal $x(n)$.

Meanwhile, the minimum Mean Square Error in the estimate of $d(n)$ may be evaluated from equation (5.2) as follows

$$\begin{aligned} \xi = E\{|e(n)|^2\} &= E\left\{e(n)\left[d(n) - \sum_{l=0}^{p-1} w(l)x(n-l)\right]^*\right\} \\ &= E\{e(n)d^*(n)\} - \sum_{l=0}^{p-1} w^*(l)E\{e(n)x^*(n-l)\} \end{aligned} \tag{5.10}$$

Recall that if $w(k)$ is the solution to the Wiener-Hopf equations, then it follows from equation (5.5) that $E\{e(n)x^*(n-k)\} = 0$. Therefore, the second term in equation (5.10) is equal to zero and

$$\xi_{\min} = E\{e(n)d^*(n)\} = E\left\{\left[d(n) - \sum_{l=0}^{p-1} w(l)x(n-l)\right]d^*(n)\right\}$$

Finally, substituting for the expected values, we have

$$\xi_{\min} = r_d(0) - \sum_{l=0}^{p-1} w(l)r_{dx}^*(l) \tag{5.11}$$

expressed in vector form as

$$\xi_{\min} = r_d(0) - \boldsymbol{r}_{dx}^{\mathrm{H}}\boldsymbol{w}. \tag{5.12}$$

Also, since we have derived equation (5.9), therefore $\boldsymbol{w} = \boldsymbol{R}_x^{-1}\boldsymbol{r}_{dx}$, the minimum error can be written as

$$\xi_{\min} = r_d(0) - \boldsymbol{r}_{dx}^{\mathrm{H}}\boldsymbol{R}_x^{-1}\boldsymbol{r}_{dx} \tag{5.13}$$

the FIR Wiener filtering equations are summarized in Table 5.1.

Table 5.1 The Wiener-Hopf Equations for the FIR Wiener Filter

Wiener-Hopf equations
$\sum_{l=0}^{p-1} w(l)r_x(k-l) = r_{dx}(k), k = 0,1,\cdots,p-1$
Correlations
$r_x(k) = E\{x(n)x^*(n-k)\}$ $r_{dx}(k) = E\{d(n)x^*(n-k)\}$
Minimum Error
$\xi_{\min} = r_d(0) - \sum_{l=0}^{p-1} w(l)r_{dx}^*(l)$

5.2.1 Filtering

In this problem, we would like to estimate $d(n)$ from a noise corrupted observation

$$x(n) = d(n) + v(n)$$

where $v(n)$ is assumed to be the noise, and it is assumed to have zero mean and to be uncorrelated with $d(n)$. Therefore, use the results in the previous section that $E\{d(n)v^*(n-k)\} = 0$ and the cross-correlation between $d(n)$ and $x(n)$ is

$$r_{dx}(k) = E\{d(n)x^*(n-k)\} = E\{d(n)d^*(n-k)\} = r_d(k) \tag{5.14}$$

When

$$\begin{aligned} r_x(k) &= E\{x(n+k)x^*(n)\} = E\{[d(n+k) + v(n+k)][d(n) + v(n]^*\} \\ &= r_d(k) + r_v(k) \end{aligned} \tag{5.15}$$

with the premise assumption that $v(n)$ and $d(n)$ are uncorrelated, it follows that $\boldsymbol{R}_x = \boldsymbol{R}_d + \boldsymbol{R}_v$, and the Wiener-Hopf equations become

$$[\boldsymbol{R}_d + \boldsymbol{R}_v]\boldsymbol{w} = \boldsymbol{r}_d, \tag{5.16}$$

where $\boldsymbol{R}_d$ is the autocorrelation matrix for $d(n)$, $\boldsymbol{R}_v$ the autocorrelation matrix for $v(n)$, and

$\boldsymbol{r}_d=[r_d(0),\cdots,r_d(p-1)]^{\mathrm{T}}$.

5.2.2 Linear Prediction

Linear prediction is an important problem in the field of signal processing research. Under the assumption of a noise-free observation environment. We want to predict the value $x(n+1)$ by linearly combining the current and previous values of $x(n)$. Thus, an FIR linear predictor of the order $p-1$ has the form

$$\hat{x}(n+1)=\sum_{k=0}^{p-1} w(k)x(n-k)$$

where $w(k)$ are the coefficients of the prediction filter. By setting $d(n)=x(n+1)$, the linear prediction problem can be included in the context of the Wiener filter problem. To set up the Wiener-Hopf equations, we need to evaluate the cross-correlation between $d(n)$ and $x(n)$. Thanks to the relationship of

$$r_{dx}(k)=E\{d(n)x^*(n-k)\}=E\{x(n+1)x^*(n-k)\}=r_x(k+1)$$

the Wiener-Hopf equations for the optimum linear predictor are

$$\begin{bmatrix} r_x(0) & r_x^*(1) & r_x^*(2) & \cdots & r_x^*(p-1) \\ r_x(1) & r_x(0) & r_x^*(1) & \cdots & r_x^*(p-2) \\ r_x(2) & r_x(1) & r_x(0) & \cdots & r_x^*(p-3) \\ \vdots & \vdots & \vdots & \ddots & \vdots \\ r_x(p-1) & r_x(p-2) & r_x(p-3) & \cdots & r_x(0) \end{bmatrix} \begin{bmatrix} w(0) \\ w(1) \\ w(2) \\ \vdots \\ w(p-1) \end{bmatrix} = \begin{bmatrix} r_x(1) \\ r_x(2) \\ r_x(3) \\ \vdots \\ r_x(p) \end{bmatrix} \tag{5.17}$$

and the Mean-Square Error is

$$\xi_{\min}=r_x(0)-\sum_{k=0}^{p-1} w(k)r_x^*(k+1)$$

Up to this point, the above derivation and results are based on the assumption of noise-free observation $x(n)$. The reality is that the linear predictor has to perform signal prediction in the presence of noisy observations. With the input to the Wiener filter given by

$$y(n)=x(n)+v(n)$$

We wish to design a filter that can estimate $x(n+1)$ in terms of a linear combination of p previous values of $y(n)$

$$\hat{x}(n+1)=\sum_{k=0}^{p-1} w(k)y(n-k)=\sum_{k=0}^{p-1} w(k)[x(n-k)+v(n-k)]$$

with the Wiener-Hopf equations in matrix form

$$\boldsymbol{R}_y\boldsymbol{w}=\boldsymbol{r}_{dy}$$

Assuming that the noise $v(n)$ is uncorrelated with the signal $x(n)$, the autocorrelation matrix for $y(n)$ is $r_y(k)=r_x(k)+r_v(k)$. Thus the only difference between linear prediction with and without noise is in the autocorrelation matrix for the input signal, where in the case of noise that is uncorrelated with $x(n)$. Based on the assumption of uncorrelation, we replace $\boldsymbol{R}_x$ with $\boldsymbol{R}_y=\boldsymbol{R}_x+\boldsymbol{R}_v$.

The linear prediction $x(n+1)$ at time $n+1$ can be summarized as a single-step linear prediction problem, which can be further developed as a multi-step linear prediction problem. Or single-step prediction $x(n+1)$ can be considered as a special case of the multi-step prediction problem $x(n+\alpha)$ with $\alpha=1$. Predicting $x(n+\alpha)$ requires that the current time $x(n)$ and the previous time have a total of p orders

$$\hat{x}(n+\alpha)=\sum_{k=0}^{p-1} w(k)x(n-k), \quad \alpha \geqslant 1$$

Compared to the single-step predictor, the Wiener-Hopf equations required for the multi-step predictor changes only at the cross-correlation vector $\boldsymbol{r}_{dx}$. In multi-step prediction problems, since $d(n)=x(n+\alpha)$, then

$$r_{dx}(k)=E\{d(n)x^*(n-k)\}=E\{x(n+\alpha)x^*(n-k)\}=r_x(\alpha+k)$$

and the Wiener-Hopf equations become

$$\begin{bmatrix} r_x(0) & r_x^*(1) & r_x^*(2) & \cdots & r_x^*(p-1) \\ r_x(1) & r_x(0) & r_x^*(1) & \cdots & r_x^*(p-2) \\ r_x(2) & r_x(1) & r_x(0) & \cdots & r_x^*(p-3) \\ \vdots & \vdots & \vdots & \ddots & \vdots \\ r_x(p-1) & r_x(p-2) & r_x(p-3) & \cdots & r_x(0) \end{bmatrix} \begin{bmatrix} w(0) \\ w(1) \\ w(2) \\ \vdots \\ w(p-1) \end{bmatrix} = \begin{bmatrix} r_x(\alpha) \\ r_x(\alpha+1) \\ r_x(\alpha+2) \\ \vdots \\ r_x(\alpha+p-1) \end{bmatrix}. \tag{5.18}$$

Equation (5.18) could be written in matrix form as

$$\boldsymbol{R}_x\boldsymbol{w}=\boldsymbol{r}_\alpha$$

where $\boldsymbol{r}_\alpha$ is the autocorrelations vector beginning with $r_x(\alpha)$. Finally, the minimum Mean Square Error is

$$\xi_{\min}=r_x(0)-\sum_{k=0}^{p-1}\omega(k)r_x^*(k+\alpha)=r_x(0)-\boldsymbol{r}_\alpha^{\mathrm{H}}\boldsymbol{w} \tag{5.19}$$

5.2.3 Noise Cancellation

Noise cancellation is an another important application of Wiener filtering. Consistent with the goal of the filtering problem, we want to estimate the signal $d(n)$ from the observations disturbed by noise, in particular when the observations are of the following form

$$x(n)=d(n)+v_1(n)$$

where $v_1(n)$ represents the inevitable additive noise. Unlike the filtering problem which requires that the autocorrelation of the noise be known, in the noise canceller this information is obtained from a secondary sensor that is placed within the noise field. Although the noise $v_2(n)$ measured by this secondary sensor will be correlated with $v_1(n)$, the two processes will not be equal.

Since the signals are acquired by two sensors separately, $d(n)$ cannot be estimated by simply subtracting $v_2(n)$ from $x(n)$. A more suitable approach is to use a Wiener filter as a noise canceller, which is designed to estimate the noise $v_1(n)$ from the signal received by the secondary sensor to obtain an estimate $\hat{v}_1(n)$. It is then subtracted from the primary signal $x(n)$, to form an estimate of $d(n)$ of the following form

$$\hat{d}(n) = x(n) - \hat{v}_1(n)$$

Using $v_2(n)$ as the input to the Wiener filter to estimate the noise $v_1(n)$, the Wiener-Hopf equations are

$$\boldsymbol{R}_{v_2}\boldsymbol{w} = \boldsymbol{r}_{v_1v_2}$$

where $\boldsymbol{R}_{v_2}$ is the autocorrelation matrix of $v_2(n)$ and $\boldsymbol{r}_{v_1v_2}$ is the vector of cross-correlations between $v_1(n)$ and $v_2(n)$. For the calculation of the cross-correlation between $v_1(n)$ and $v_2(n)$, we have

$$\begin{aligned} r_{v_1v_2}(k) &= E\{v_1(n)v_2^*(n-k)\} = E\{[x(n) - d(n)]v_2^*(n-k)\} \\ &= E\{x(n)v_2^*(n-k)\} - E\{d(n)v_2^*(n-k)\} \end{aligned} \tag{5.20}$$

If we assume that $v_2(n)$ is uncorrelated with $d(n)$, then the second term of equation (5.20) equals to zero and the cross-correlation becomes

$$r_{v_1v_2}(k) = E\{x(n)v_2^*(n-k)\} = r_{xv_2}(k) \tag{5.21}$$

Finally, the Wiener-Hopf equations are

$$\boldsymbol{R}_{v_2}\boldsymbol{w} = \boldsymbol{r}_{xv_2} \tag{5.22}$$

5.3 The IIR Wiener Filter

The problems discussed in this section are in the scope of the FIR Wiener filter design problem. Now we consider the design of an IIR digital Wiener filter. As in the case of the FIR Wiener filter, when givena $x(n)$ our goal is to design a filter $h(n)$ that produces an output $y(n) = x(n) * h(n)$ that approximates a desired process $d(n)$ in the mean-square sense. Although the FIR and IIR Wiener filters have the same formulation of this problem, an important difference is that only a finite number of filter coefficients must be determined for the FIR Wiener filter whereas there is an infinite number of unknowns for the IIR Wiener filter.

In this section, we develop a discussion on the causality of IIR. First, in Section 5.3.1, we will solve the Wiener filter design problem with no constraints on the solution. Subsequently, we will find that the optimum filter is noncausal and unrealizable. Then, in Section 5.3.2, we will constrain the solution to be causal by forcing $h(n)$ to be zero for $n \leqslant 0$. And for the noncausal Wiener filter, we will find a simple closed-form expression for the frequency response. For the causal Wiener filter, we will only specify the system function implicitly in terms of a spectral factorization.

5.3.1 Noncausal IIR Wiener Filter

For solving a noncausal IIR Wiener filter without constraints, the problem is to find the unit impulse response $h(n)$ of the IIR filter

$$H(z) = \sum_{n=-\infty}^{\infty} h(n)z^{-n}$$

that minimizes $\xi = E\{|e(n)|^2\}$, where $e(n)$ is the error between $d(n)$ and the estimate of the Wiener filter $\hat{d}(n)$, written as

$$e(n) = d(n) - \hat{d}(n) = d(n) - \sum_{l=-\infty}^{\infty} h(l)x(n-l) \tag{5.23}$$

The solution is the same way we solved the FIR Wiener filtering problem, by differentiating ξ with concerning to $h^*(k)$ for each k and setting the derivatives equal to zero

$$\frac{\partial \xi}{\partial h^*(k)} = \frac{\partial E\{e(n)e^*(n)\}}{\partial h^*(k)} = -E\{e(n)x^*(n-k)\} = 0, \ -\infty < k < \infty$$

and is equivalent to

$$E\{e(n)x^*(n-k)\} = 0, \ -\infty < k < \infty \tag{5.24}$$

As $e(n)$ in equation (5.23) is brought into equation (5.24), there is

$$\sum_{l=-\infty}^{\infty} h(l)E\{x(n-l)x^*(n-k)\} = E\{d(n)x^*(n-k)\}, \ -\infty < k < \infty \tag{5.25}$$

The expectation on the left side is the autocorrelation of $x(n)$, and the expectation on the right side is the cross-correlation between $x(n)$ and $d(n)$. Thus, equation (5.25) may be written as

$$\sum_{l=-\infty}^{\infty} h(l)r_x(k-l) = r_{dx}(k), \ -\infty < k < \infty \tag{5.26}$$

which are the Wiener-Hopf equations of the noncausal IIR Wiener filter. Comparing the Wiener-Hopf equations for the FIR Wiener filter and the noncausal IIR Wiener filter, (equations (5.7) and (5.26)), one can find that the difference lies in the limits on the sum and the range of values over which the equations must hold. Although equation (5.26) corresponds to a set of linear equations with an infinite number of unknowns, it may seem unsolvable. However, if the left side is written as the convolution of $h(k)$ with $r_x(k)$, as in equation (5.27), then the problem can be easily handled by starting from the frequency domain.

$$h(k) * r_x(k) = r_{dx}(k) \tag{5.27}$$

In the frequency domain, equation (5.27) becomes

$$H(e^{j\omega})P_x(e^{j\omega}) = P_{dx}(e^{j\omega}) \tag{5.28}$$

Therefore, the frequency response of the IIR Wiener filter is

$$H(e^{j\omega}) = \frac{P_{dx}(e^{j\omega})}{P_x(e^{j\omega})} \tag{5.29}$$

and the system function is

$$H(z) = \frac{P_{dx}(z)}{P_x(z)} \tag{5.30}$$

where the denominator is a power spectral density $P_x(z)$, and the numerator $P_{dx}(z)$, is a cross-power spectral density.

With the above results, we now evaluate the Mean Square Error. Following the steps that were taken for an FIR Wiener filter, the Mean-Square Error is

$$\xi_{\min} = r_d(0) - \sum_{l=-\infty}^{\infty} h(l)r_{dx}^*(l) \tag{5.31}$$

Using Parseval's theorem, we transform the error representation into a form in the frequency domain as follows

$$\xi_{\min} = r_d(0) - \frac{1}{2\pi}\int_{-\pi}^{\pi} H(e^{j\omega})P_{dx}^*(e^{j\omega})d\omega \tag{5.32}$$

Since

$$r_d(0) = \frac{1}{2\pi}\int_{-\pi}^{\pi} P_d(e^{j\omega})d\omega$$

then equation (5.32) can be further written as

$$\xi_{\min} = \frac{1}{2\pi}\int_{-\pi}^{\pi} [P_d(e^{j\omega}) - H(e^{j\omega})P_{dx}^*(e^{j\omega})]d\omega \tag{5.33}$$

The error may also be expressed in terms of the complex variable z as follows:

$$\xi_{\min} = \frac{1}{2\pi j}\oint_C \left[P_d(z) - H(z)P_{dx}^*\left(\frac{1}{z^*}\right)\right]z^{-1}dz \tag{5.34}$$

where the contour C can be considered as the unit circle. The noncausal Wiener filtering equations are summarized in Table 5.2.

Table 5.2 The Frequency Response for a Noncausal Wiener Filter

Wiener-Hopf equations
$\sum_{l=-\infty}^{\infty} h(l)r_x(k-l) = r_{dx}(k), \quad -\infty < k < \infty$
Correlations
$r_x(k) = E\{x(n)x^*(n-k)\}$ $r_{dx}(k) = E\{d(n)x^*(n-k)\}$
Minimum Error
$\xi_{\min} = r_d(0) - \sum_{l=-\infty}^{\infty} h(l)r_{dx}^*(l) = \frac{1}{2\pi}\int_{-\pi}^{\pi}[P_d(e^{j\omega}) - H(e^{j\omega})P_{dx}^*(e^{j\omega})]d\omega$

At the end of this subsection, We derive the Wiener smoothing filter for producing the minimum mean-square estimate of a process $d(n)$ using the noisy observations

$$x(n) = d(n) + v(n)$$

Assuming that $d(n)$ and $v(n)$ are uncorrelated zero mean random processes, the autocorrelation of $x(n)$ is $r_x(k) = r_d(k) + r_v(k)$, and the power spectrum is

$$P_x(e^{j\omega}) = P_d(e^{j\omega}) + P_v(e^{j\omega})$$

Furthermore, the cross-correlation $r_{dx}(k)$ is

$$r_{dx}(k) = E\{d(n)x^*(n-k)\} = E\{d(n)d^*(n-k)\} + E\{d(n)v^*(n-k)\} = r_d(k)$$

Therefore,

$$P_{dx}(e^{j\omega}) = P_d(e^{j\omega})$$

and the IIR Wiener smoothing filter is

$$H(e^{j\omega}) = \frac{P_d(e^{j\omega})}{P_d(e^{j\omega}) + P_v(e^{j\omega})} \tag{5.35}$$

For those values of ω in which $P_d(e^{j\omega}) \gg P_v(e^{j\omega})$, the signal-to-noise ratio is high and $|H(e^{j\omega})| \approx 1$. Therefore, in these frequency bands the components have little attenuation during their passage through the filter. On the other hand, for those values of ω where the signal-to-noise ratio is small, $P_d(e^{j\omega}) \ll P_v(e^{j\omega})$, the frequency response is small, and $|H(e^{j\omega})| \approx 0$. Thus, on those frequency bands where the noise dominates, $H(e^{j\omega})$ is small to filter out or suppress the noise.

Finally, since $P_{dx}(e^{j\omega}) = P_d(e^{j\omega})$, if we evaluate the Mean Square Error using equation (5.33) and use the fact that $P_d(e^{j\omega})$ is real, we have

$$\xi_{\min} = \frac{1}{2\pi}\int_{-\pi}^{\pi}[P_d(e^{j\omega}) - H(e^{j\omega})P_{dx}^*(e^{j\omega})]d\omega = \frac{1}{2\pi}\int_{-\pi}^{\pi}P_d(e^{j\omega})[1 - H(e^{j\omega})]d\omega$$

Substituting equation (5.35) for the frequency response of the Wiener smoothing filter we find that the minimum Mean Square Error is

$$\begin{aligned}\xi_{\min} &= \frac{1}{2\pi}\int_{-\pi}^{\pi}P_d(e^{j\omega})\left[1 - \frac{P_d(e^{j\omega})}{P_d(e^{j\omega}) + P_v(e^{j\omega})}\right]d\omega \\ &= \frac{1}{2\pi}\int_{-\pi}^{\pi}P_d(e^{j\omega})\frac{P_v(e^{j\omega})}{P_d(e^{j\omega}) + P_v(e^{j\omega})}d\omega = \frac{1}{2\pi}\int_{-\pi}^{\pi}H(e^{j\omega})P_v(e^{j\omega})d\omega\end{aligned} \tag{5.36}$$

if expressed in the z -domain, becomes

$$\xi_{\min} = \frac{1}{2\pi j}\oint_C H(z)P_v(z)z^{-1}dz \tag{5.37}$$

5.3.2 The Causal IIR Wiener Filter

In this subsection, we derive the design of the case where the IIR digital Wiener filter is constrained to be causal. In this case, the unit impulse response is zero for $n < 0$ and the estimate of $d(n)$ takes the form of

$$\hat{d}(n) = x(n) * h(n) = \sum_{k=0}^{\infty} h(k)x(n-k)$$

To find the filter coefficients that minimize the Mean Square Error, we proceed exactly in the same way that we did for the noncausal Wiener filter. Specifically, differentiating ξ with concerning to $h^*(k)$ for $k \geqslant 0$ and setting the derivatives to zero we find

$$\sum_{l=0}^{\infty} h(l)r_x(k-l) = r_{dx}(k), 0 \leqslant k < \infty \tag{5.38}$$

which are the Wiener-Hopf equations for the causal IIR Wiener filter.

To solve the Wiener-Hopf equations, we begin by looking at the special case in which the input to the filter is a white noise whose variance is equal to 1, $\epsilon(n)$. Denoting the coefficients of the Wiener filter by $g(n)$, the Wiener-Hopf equations are

$$\sum_{l=0}^{\infty} g(l)r_\epsilon(k-l) = r_{d\epsilon}(k), 0 \leqslant k < \infty \tag{5.39}$$

With $r_\epsilon(k-l) = \delta(k-l)$, the left side of equation (5.39) reduces to $g(k)$. Therefore, $g(k) = r_{d\epsilon}(k)$ for $k \geqslant 0$ and, since the Wiener filter is causal, $g(k) = 0$ for $k < 0$. Thus, the causal Wiener filter for a white noise input $\epsilon(n)$ is

$$g(n) = r_{d\epsilon}(n)u(n) \tag{5.40}$$

where $u(n)$ is the unit step function. We will express this solution in the z-domain as follows

$$G(z) = [P_{d\epsilon}(z)]_+ \tag{5.41}$$

where the subscript "+" is used to indicate the "positive-time part" of the sequence whose z-transform is contained within the brackets.

In a typical Wiener filtering application, it is unlikely that the input to the Wiener filter will be white noise. Therefore, suppose that $x(n)$ is a random process with a rational power spectrum that has no poles or zeros on the unit circle. We may then perform a spectral factorization and write $P_x(z)$ as follows

$$P_x(z) = \sigma_0^2 Q(z) Q^*(1/z^*) \tag{5.42}$$

where $Q(z)$ is minimum phase and of the form

$$Q(z) = 1 + q(1)z^{-1} + q(2)z^{-2} + \cdots = \frac{N(z)}{D(z)}$$

with $N(z)$ and $D(z)$ minimum phase monic polynomials. If $x(n)$ is filtered with a filter having a system function of the form (see Fig. 5.1)

$$F(z) = \frac{1}{\sigma_0 Q(z)} \tag{5.43}$$

then the power spectrum of the output process, $\epsilon(n)$, will be

$$P_\epsilon(z) = P_x(z)F(z)F^*(1/z^*) = 1$$

Therefore, $\epsilon(n)$ is white noise and $F(z)$ is referred to as a whitening filter. Note that since $Q(z)$ is minimum phase, then $F(z)$ is stable and causal and has a stable and causal inverse, $F^{-1}(z)$. As a result, $x(n)$ may be recovered from $\epsilon(n)$ by filtering with the inverse filter, $F^{-1}(z)$. In other words, there is no loss of information in the linear transformation that produces the white noise process from $x(n)$.

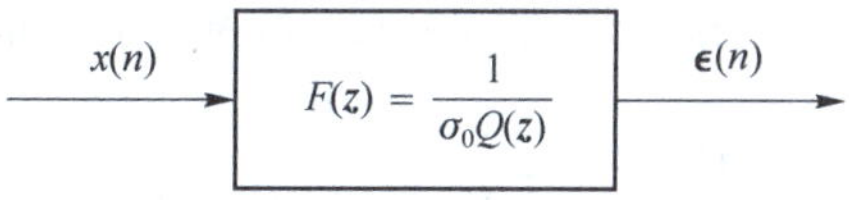

Figure 5.1 A whitening filter that produces white noise with power spectrum $P_\epsilon(e^{j\omega}) = 1$ when the input $x(n)$ has a power spectrum $P_x(z) = \sigma_0^2 Q(z)Q^*(1/z^*)$

With this background, we are now in a position to derive the optimum causal Wiener filter when the input to the filter $x(n)$ has a rational power spectrum. Let $H(z)$ be the causal Wiener filter that produces the minimum mean-square estimate of $d(n)$ from $x(n)$, and suppose that $x(n)$ is filtered with a cascade of three filters, $F(z)$, $F^{-1}(z)$, and $H(z)$ as shown in Figure 5.2, where $F(z)$ is the causal whitening filter for $x(n)$ and $F^{-1}(z)$ is the causal inverse. The cascade $G(z) = F^{-1}(z)H(z)$ is the causal Wiener filter that produces the minimum mean-square estimate of $d(n)$ from the white noise process $\epsilon(n)$. The causality of $G(z)$ follows from the fact that both $F^{-1}(z)$ and $H(z)$ are causal.

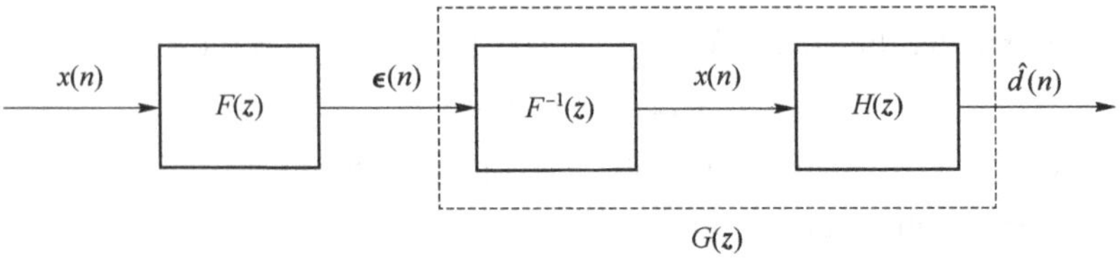

Figure 5. 2 A causal Wiener filter $G(z)$ operating on a whitened input $\epsilon(n)$ with $H(z)$ the causal Wiener filter for estimating $d(n)$ from $x(n)$

With $\epsilon(n)$ a white noise process, we see from equation (5. 41) that $G(z)$, the causal IIR Wiener filter for estimating $d(n)$ from $\epsilon(n)$, is $G(z) = [P_{d\epsilon}(z)]_+$. Since $\epsilon(n)$ is formed by filtering $x(n)$ with the whitening filter $f(n)$, then the cross-correlation between $d(n)$ and $\epsilon(n)$ is

$$r_{d\epsilon}(k) = E\{d(n)\epsilon^*(n-k)\}$$

$$= E\left\{d(n)\left[\sum_{l=-\infty}^{\infty} f(l)x(n-k-l)\right]^*\right\} = \sum_{l=-\infty}^{\infty} f^*(l)r_{dx}(k+l) \tag{5.44}$$

Therefore, the cross-power spectral density $P_{d\epsilon}(z)$ is

$$P_{d\epsilon}(z) = F^*(1/z^*)P_{dx}(z) = \frac{P_{dx}(z)}{\sigma_0 Q^*(1/z^*)}$$

and the causal Wiener filter for estimating $d(n)$ from $\epsilon(n)$ is

$$G(z) = \frac{1}{\sigma_0}\left[\frac{P_{dx}(z)}{Q^*\left(\frac{1}{z^*}\right)}\right]_+ \tag{5.45}$$

Thus, combining equations (5. 43) and (5. 45) lead to the desired solution

$$H(z) = \frac{1}{\sigma_0^2 Q(z)}\left[\frac{P_{dx}(z)}{Q^*(1/z^*)}\right]_+ \tag{5.46}$$

In the case of real processes, $h(n)$ is real and the causal Wiener filter takes the form

$$H(z) = \frac{1}{\sigma_0^2 Q(z)}\left[\frac{P_{dx}(z)}{Q(z^{-1})}\right]_+ \tag{5.47}$$

Finally, as with the noncausal IIR Wiener filter, the Mean Square Error for the causal IIR Wiener filter is

$$\xi_{\min} = r_d(0) - \sum_{l=0}^{\infty} h(l)r_{dx}^*(l) \tag{5.48}$$

where the sum extends only over the interval $0 \leqslant l < \infty$ since $h(l) = 0$ for $l < 0$. In the frequency domain, this error may be written as

$$\xi_{\min} = \frac{1}{2\pi}\int_{-\pi}^{\pi} [P_d(e^{j\omega}) - H(e^{j\omega})P_{dx}^*(e^{j\omega})]d\omega \tag{5.49}$$

or, equivalently

$$\xi_{\min} = \frac{1}{2\pi j}\oint_C \left[P_d(z) - H(z)P_{dx}^*\left(\frac{1}{z^*}\right)\right]z^{-1}dz \tag{5.50}$$

The causal Wiener filtering equations are summarized in Table 5.3

Table 5.3 The System Function and Minimum Error for a Causal Wiener Filter

System function
$$H(z) = \frac{1}{\sigma_0^2 Q(z)}\left[\frac{P_{dx}(z)}{Q^*(1/z^*)}\right]_+$$
Spectral Factorization
$$P_x(z) = \sigma_0^2 Q(z) Q^*(1/z^*)$$
Minimum Error
$$\xi_{\min} = r_d(0) - \sum_{l=0}^{\infty} h(l) r_{dx}^*(l) = \frac{1}{2\pi}\int_{-\pi}^{\pi}[P_d(e^{j\omega}) - H(e^{j\omega})P_{dx}^*(e^{j\omega})]d\omega$$

So far, under the condition that the first-order and second-order prior knowledge is known, we have completed the exploration of the FIR and IIR Wiener filtering problems. And when the conditions are more realistic, it is difficult for us to grasp these statistical properties before obtaining the observations. How to design the filter under such assumptions becomes the issue to be discussed in the next subsection.

5.4 Least-Squares Filter

In this section, we discuss the related aspects of Least Squares, including its principles and the problem of solving it when designing filters. The principles of Least Squares are widely used in practice because a prior knowledge, especially second order moments, is rarely available in advance.

First, we will give an introduction to the properties of Least Squares, and Section 5.4.2 discusses Least Squares FIR filters and predictors. In Section 5.4.3, we solve the LSE normal equation and introduce the singular value decomposition (SVD) that can be used to compute LSE estimates.

5.4.1 The Least Square Principle

The problem of filter design under the MMSE criterion requires a priori knowledge of second-order statistics. However, such statistical information is not available in most practical applications, and we would like to design filters using only the inputs and observations available.

To achieve this, in this subsection, we temporarily abandon the MMSE criterion and instead design an optimal filter by minimizing the error between the desired signal $d(n)$ and its estimate $\hat{d}(n)$, a method known as least squared error (LSE) estimation.

We start with the derivation of the general linear LS filter. For a linear combiner of p order, assume that the input is $x(n)$, and the output denoted by $\hat{d}(n)$, with the form of

$$\hat{d}(n) = \sum_{k=0}^{p-1} x_k^*(n) w_k = \boldsymbol{x}^{\mathrm{H}}(n)\boldsymbol{w}. \tag{5.51}$$

The error between it and the desired signal is defined as

$$e(n) = d(n) - \hat{d}(n) = d(n) - x^{\mathrm{H}}(n)\boldsymbol{w}, \tag{5.52}$$

and for solving the coefficients $w(n)$, this is achieved by minimizing the sum of squared errors, which is written as

$$E = \sum_{n=0}^{N-1} |e(n)|^2. \tag{5.53}$$

It can also be considered as the energy of error, and it is worth noting that the sum of N errors is taken here to make the coefficient stable when receiving N inputs. Since it has become a squared term. Equation (5.52) can also be written in matrix form as in the previous reasoning for the Wiener-Hopf equations, each term in the equation is lifted to vector or matrix form, and eventually equation (5.52) becomes

$$\begin{bmatrix} e(0) \\ e(1) \\ \vdots \\ e(N-1) \end{bmatrix} = \begin{bmatrix} d(0) \\ d(1) \\ \vdots \\ d(N-1) \end{bmatrix} - \begin{bmatrix} x_0^*(0) & x_1^*(0) & \cdots & x_{p-1}^*(0) \\ x_0^*(1) & x_1^*(1) & \cdots & x_{p-1}^*(1) \\ \vdots & \vdots & \ddots & \vdots \\ x_0^*(N-1) & x_1^*(N-1) & \cdots & x_{p-1}^*(N-1) \end{bmatrix} \begin{bmatrix} w_0 \\ w_1 \\ \vdots \\ w_{p-1} \end{bmatrix} \tag{5.54}$$

or

$$\boldsymbol{e} = \boldsymbol{d} - \boldsymbol{X}\boldsymbol{w}. \tag{5.55}$$

It follows that the signal processing under the LSE criterion can also be performed in a batch mode, which involves the processing of N inputs at a time, corresponding to N momentary outputs, and the length taken by each input is related to the order of the combiner. And when we pursue $\boldsymbol{e}=\boldsymbol{0}$ under this criterion, it is equivalent to solving a matrix of N equations with p unknown parameters. If $N=p$, then equation (5.54) usually yields a unique solution. If $N > p$, then it means an overdetermined system of equations that has no solution. Or when $N < p$, we have an underdetermined system that has an infinite set of solutions. However for the latter two cases, we can also obtain unique Least Squares solutions by some methods.

Next we will first deal with the design of FIR filters on the overdetermined system, since they play a very important role in practical applications.

5.4.2 Least Square FIR Filters

We now apply the LSE criterion to the design of FIR filters, processed in a way that is closely related to Wiener filtering. In equation (5.52) we have defined that the error between $d(n)$ and its estimate is

$$e(n) = d(n) - \sum_{k=0}^{p-1} x(n-k) w_k = d(n) - \boldsymbol{x}(n)\boldsymbol{w},$$

where the $\boldsymbol{w}$ is the filter coefficient vector that we try to determine, and further we extend it to matrix form as

$$e = d - Xw.$$

After eliminating its sign through equation (5.53), we can proceed to push it to the minimum, thus obtaining the optimal weight vector based on observations and desired signals.

$$\begin{aligned} E &= \sum_{n=0}^{N-1} |e(n)|^2 = e^{H} e \\ &= (d - Xw)^{H}(d - Xw) \\ &= d^{H} d - d^{H} X w - w^{H} X^{H} d + w^{H} X^{H} X w \end{aligned} \tag{5.56}$$

Here we define $X^{H}X = R_X$ and $X^{H}d = r_{Xd}$. However, unlike the real second order statistics of $x(n)$ or $d(n)$, they only represent the currently processed $X(n)$ and $d(n)$ with no statistical properties. With $d^{H}d = E_d$, equation (5.56) becomes

$$E = E_d - r_{Xd}^{H} w - w^{H} r_{Xd} + w^{H} R_X w. \tag{5.57}$$

We then differentiate w in equation (5.57) to obtain the minimum value

$$\begin{aligned} \frac{\partial E}{\partial w} &= - r_{Xd}^{H} - r_{Xd}^{H} + 2 w_{LS}^{H} R_X = 0 \\ &\Rightarrow r_{Xd}^{H} = w_{LS}^{H} R_X \end{aligned} , \tag{5.58}$$

thus

$$\begin{aligned} r_{Xd}^{H} &= w_{LS}^{H} R_X \Rightarrow X^{H} X w_{LS} = Xd, \\ w_{LS} &= (X^{H} X)^{-1} Xd, \end{aligned} \tag{5.59}$$

this leads to the weight vector under LSM criterion. At this point, we have solved the Least Squares FIR filter problem. The significant difference compared to the previously mentioned MMSE criterion is that it replaces the prior knowledge with the $X^{H}X$, thus greatly reducing the difficulty in practical using.

Another point is that usually such a way of designing an FIR filter is based on the condition of an overdetermined system of equations, where $N > p$. In the next subsections, we talk about the method that deals with the problem of solving the weights under the system of underdetermined equations.

5.4.3 Least Square Computations

In this section, we discuss how to solve for w in the case that $X \in \mathbb{R}^{N \times p}$ when $N < p$. It leads to the question of how we can determine the value of the unknowns as much as possible when the number of unknowns is greater than the number of equations. Here we focus on the singular value decomposition (SVD) method, which is the most widely used method for solving systems of rank-deficient equations, In fact, in applications of singular value analysis, it is often necessary to approximate a matrix with noise or perturbation by a matrix of low rank. Whereas in signal processing and system theory, the most common systems of linear equations such as $Ab = C$ are overdetermined and rank deficient, SVD can endow them with a solution $\hat{b}$ that has minimum residuals, which is very similar to the problem we are going to solve about minimizing $\|d - Xw\|_2$.

First, we introduce the analysis process of SVD. Assuming that $\boldsymbol{A}$ is a $N \times p$ matrix or rank k. Then, there exists a $N \times N$ orthogonal matrix $\boldsymbol{U}$ and a $p \times p$ orthogonal matrix $\boldsymbol{V}$, a $N \times p$ diagonal matrix $\sum$, and the four have a relationship that

$$\boldsymbol{A} = \boldsymbol{U} \sum \boldsymbol{V}^{\mathrm{H}}, \tag{5.60}$$

$\sum$ defines as

$$\sum = \begin{bmatrix} \boldsymbol{S} & \boldsymbol{0} \\ \boldsymbol{0} & \boldsymbol{0} \end{bmatrix}, \text{where } \boldsymbol{S} = \mathrm{diag}(\sigma_1, \sigma_2, \cdots, \sigma_k). \tag{5.61}$$

According to the matrix decomposition theory, the first k singular values $\sigma_1, \sigma_2, \cdots, \sigma_k$ are arranged in strict order from largest to smallest and $\sigma_1, \sigma_2, \cdots, \sigma_k \geqslant 0$. The remaining $N - k$ singular values are equal to 0.

The decomposed matrix $\boldsymbol{A}$ can be rewritten in the form of a regular linear combination as

$$\boldsymbol{A} = \sum_{i=1}^{k} \sigma_i \boldsymbol{u}_i \boldsymbol{v}_i^{\mathrm{H}}, \tag{5.62}$$

$\boldsymbol{u}_i$, the column vectors of $\boldsymbol{U}$, are the left singular vectors of $\boldsymbol{A}$, and $\boldsymbol{v}_i$, the column vectors of $\boldsymbol{V}$, are the right singular vectors of $\boldsymbol{A}$. It is worth noting that the singular value σ_i is the non-negative square root of the eigenvalues of $\boldsymbol{A}^{\mathrm{H}}\boldsymbol{A}$. Therefore when returning to the problem we are trying to solve above, the k singular values corresponding to $\boldsymbol{X}$ can be calculated by the process of finding the eigenvalues of $\boldsymbol{X}^{\mathrm{H}}\boldsymbol{X}$ followed by square rooting.

We are now faced with the problem of how to solve for the weight vector when $N < p$. For the problem in equation (5.55) that $\boldsymbol{e} = \boldsymbol{d} - \boldsymbol{X}\boldsymbol{w}$, by using SVD method, we decompose $\boldsymbol{X}$ according to equation (5.60), and obtain

$$\boldsymbol{e} = \boldsymbol{d} - \boldsymbol{U} \sum \boldsymbol{V}^{\mathrm{H}} \boldsymbol{w}, \tag{5.63}$$

and $\boldsymbol{U}$ is the orthogonal matrix, thus $\|\boldsymbol{e}\|^2 = \|\boldsymbol{U}^{\mathrm{H}}\boldsymbol{e}\|^2$, and the effect of minimizing the right side items is equivalent to minimizing $\|\boldsymbol{e}\|^2$, hence equation becomes

$$\boldsymbol{U}^{\mathrm{H}}\boldsymbol{e} = \boldsymbol{U}^{\mathrm{H}}\boldsymbol{d} - \sum \boldsymbol{V}^{\mathrm{H}}\boldsymbol{w}. \tag{5.64}$$

We first partition the orthogonal matrices $\boldsymbol{U}$ and $\boldsymbol{V}$ as

$$\begin{aligned} \boldsymbol{U} &= [\boldsymbol{U}_1 \quad \boldsymbol{U}_2] \\ \boldsymbol{V} &= [\boldsymbol{V}_1 \quad \boldsymbol{V}_2] \end{aligned}, \tag{5.65}$$

and following the position of the elements of $\sum$ in equation (5.61), it follows that $\boldsymbol{U}_1$ and $\boldsymbol{V}_1$ contain the column vectors for the left and right singular values associated with the nonzero singular values of $\boldsymbol{X}$, $\boldsymbol{U}_2$ and $\boldsymbol{V}_2$ contain the column vectors associated with the zero singular values, respectively. We next substitute into equation (5.64), and obtain

$$\boldsymbol{U}^{\mathrm{H}}\boldsymbol{e} = \begin{bmatrix} \boldsymbol{U}_1^{\mathrm{H}} \\ \boldsymbol{U}_2^{\mathrm{H}} \end{bmatrix} \boldsymbol{d} - \begin{bmatrix} \boldsymbol{S} & \boldsymbol{0} \\ \boldsymbol{0} & \boldsymbol{0} \end{bmatrix} \begin{bmatrix} \boldsymbol{V}_1^{\mathrm{H}} \\ \boldsymbol{V}_2^{\mathrm{H}} \end{bmatrix} \boldsymbol{w} = \begin{bmatrix} \boldsymbol{U}_1^{\mathrm{H}}\boldsymbol{d} - \boldsymbol{S}\boldsymbol{V}_1^{\mathrm{H}}\boldsymbol{w} \\ \boldsymbol{U}_2^{\mathrm{H}}\boldsymbol{d} \end{bmatrix} \tag{5.66}$$

Therefore,

$$\|\boldsymbol{e}\|^2 = \|\boldsymbol{U}^{\mathrm{H}}\boldsymbol{e}\|^2 = \|\boldsymbol{U}_1^{\mathrm{H}}\boldsymbol{d} - \boldsymbol{S}\boldsymbol{V}_1^{\mathrm{H}}\boldsymbol{w}\|^2 + \|\boldsymbol{U}_2^{\mathrm{H}}\boldsymbol{d}\|^2 \tag{5.67}$$

To this point, to minimize the squared term of the error, on the one hand, in the latter term, we seek to minimize $\|\boldsymbol{U}_2^{\mathrm{H}}\boldsymbol{d}\|^2$. On the other hand, we wish to make $\|\boldsymbol{U}_1^{\mathrm{H}}\boldsymbol{d}-\boldsymbol{S}\,\boldsymbol{V}_1^{\mathrm{H}}\boldsymbol{w}\|^2$ equals to zero, which gives

$$\boldsymbol{S}\,\boldsymbol{V}_1^{\mathrm{H}}\hat{\boldsymbol{w}} = \boldsymbol{U}_1^{\mathrm{H}}\boldsymbol{d} \tag{5.68}$$

and equals to

$$\boldsymbol{V}_1^{\mathrm{H}}\hat{\boldsymbol{w}} = \boldsymbol{S}^{-1}\,\boldsymbol{U}_1^{\mathrm{H}}\boldsymbol{d} \tag{5.69}$$

From this, we derive the coefficient vector for solving the underdetermined equations with SVD method, it is expressed as

$$\hat{\boldsymbol{w}} = \sum_{i=1}^{k} \frac{\boldsymbol{u}_i^{\mathrm{H}}\boldsymbol{d}}{\sigma_i}\boldsymbol{v}_{\mathrm{i}} \tag{5.70}$$

k is the rank of $\boldsymbol{X}$, and the solution $\hat{\boldsymbol{w}}$ is unique and is the minimum norm solution to the least square problem.

The above is a derivation to apply the SVD method to the Least Squares estimation of rank-deficient signals. Another problem is to determine the rank of the observation $\boldsymbol{X}$, which can be done by examining $\boldsymbol{X}^{\mathrm{H}}\boldsymbol{X}$, the so-called autocorrelation $\boldsymbol{R}_X$, for the eigenvalues σ_i^2 is the square term of the singular values σ_i which we have already mentioned before. The eigenvalues can be used to estimate the rank of $\boldsymbol{X}$, where the larger ones are contributed by the signal, and the smaller ones are caused by the noise. This method provides a reliable estimate of the rank because usually, the energy levels of both the signal and the noise are not at the same level.

In the last section, we will return to the MMSE criterion under the Wiener filter and explore how to make signal predictions based on observations and some prior knowledge alone.

5.5 Discrete Kalman Filter

In Section 5.3.2 we considered the problem of designing a causal Wiener filter to estimate a process $d(n)$ from a set of noisy observations $x(n) = d(n) + v(n)$. The primary limitation of the solution that was derived is that it requires that $d(n)$ and $x(n)$ be jointly wide-sense stationary processes. Since most processes encountered in practice are nonstationary, this constraint limits the usefulness of the Wiener filter. Therefore, in this section we re-examine this estimation problem within the context of nonstationary processes and derive what is known as the discrete Kalman filter.

To begin, let us look briefly once again at the causal Wiener filter for estimating a process $x(n)$ from the noisy measurements

$$y(n) = x(n) + v(n)$$

we considered the specific problem of estimating an AR(1) process of the form

$$x(n) = a(1)x(n-1) + \omega(n)$$

where $\omega(n)$ and $v(n)$ are uncorrelated white noise processes. What we discovered was that the optimum linear estimate of $x(n)$, using all of the measurements, $y(k)$, for $k \leqslant n$, could be computed with a recursion of the form

$$\hat{x}(n) = a(1)\hat{x}(n-1) + K[y(n) - a(1)\hat{x}(n-1)] \tag{5.71}$$

where K is a constant, referred to as the Kalman gain, that minimizes the Mean Square Error $E\{|x(n)-\hat{x}(n)|^2\}$.

However, there are two problems with this solution that need to be addressed. First is the requirement that $x(n)$ and $y(n)$ be jointly wide-sense stationary processes. For example, equation (5.51) is not the optimum linear estimate if $x(n)$ is a nonstationary process, such as the one that is generated by the time-varying difference equation

$$x(n)=a_{n-1}(1)x(n-1)+\omega(n)$$

Nevertheless, what we will discover is that the optimum estimate may be written as

$$\hat{x}(n)=a_{n-1}(1)\hat{x}(n-1)+K(n)[y(n)-a_{n-1}(1)\hat{x}(n-1)] \tag{5.72}$$

where $K(n)$ is a suitably chosen (time-varying) gain. The second problem with the Wiener solution is that it does not allow the filter to be "turned on" at time $n=0$. In other words, implicit in the development of the causal Wiener filter is the assumption that the observations $y(k)$ are available for all $k\leqslant n$. Again, however, we will find that this problem is addressed with an estimate of the form given in equation (5.72).

Although the discussion above is only concerned with the estimation of an AR(1) process from noisy measurements, using state variables we may easily extend these results to more general processes. For example, let $x(n)$ be an AR (p) process that is generated according to the difference equation

$$x(n)=\sum_{k=1}^{p}a(k)x(n-k)+\omega(n) \tag{5.73}$$

and suppose that $x(n)$ is measured in the presence of additive noise

$$y(n)=x(n)+v(n)$$

If we let $\boldsymbol{x}(n)$ be the p-dimensional state vector

$$\boldsymbol{x}(n)=\begin{bmatrix} x(n) \\ x(n-1) \\ \vdots \\ x(n-p+1) \end{bmatrix}$$

then equation (5.73) may be written in terms of $\boldsymbol{x}(n)$ as follows

$$\boldsymbol{x}(n)=\begin{bmatrix} a(1) & a(2) & \cdots & a(p-1) & a(p) \\ 1 & 0 & \cdots & 0 & 0 \\ 0 & 1 & \cdots & 0 & 0 \\ \vdots & \vdots & \cdots & \vdots & \vdots \\ 0 & 0 & \cdots & 1 & 0 \end{bmatrix}\boldsymbol{x}(n-1)+\begin{bmatrix} 1 \\ 0 \\ 0 \\ \vdots \\ 0 \end{bmatrix}\omega(n) \tag{5.74}$$

and

$$y(n)=[1,0,\cdots,0]\boldsymbol{x}(n)+v(n) \tag{5.75}$$

Using matrix notation to simplify these equations, we have

$$\begin{aligned} \boldsymbol{x}(n)&=\boldsymbol{A}\boldsymbol{x}(n-1)+\boldsymbol{\omega}(n) \\ y(n)&=\boldsymbol{C}^{\mathrm{T}}\boldsymbol{x}(n)+\boldsymbol{v}(n) \end{aligned} \tag{5.76}$$

where A is a $p \times p$ state transition matrix, $\boldsymbol{\omega}(n) = [\omega(n), 0, \cdots, 0]^T$ is a vector noise process, and ***C* is a unit vector of length** p. As in equation (5.72) for the case of an AR(1) process, the optimum estimate of the state vector $\boldsymbol{x}(n)$, using all of the measurements up to time n, may be expressed in the form

$$\hat{\boldsymbol{x}}(n) = \boldsymbol{A}\hat{\boldsymbol{x}}(\mathrm{n}-1) + \boldsymbol{K}[y(n) - \boldsymbol{C}^T\boldsymbol{A}\hat{\boldsymbol{x}}(n-1)] \tag{5.77}$$

where $\boldsymbol{K}$ is a Kalman gain vector.

Although only applicable to stationary AR(p) processes, equation (5.56) may be easily generalized to nonstationary processes as follows. Let $\boldsymbol{x}(n)$ be a state vector of dimension p that evolves according to the difference equation

$$\boldsymbol{x}(n) = \boldsymbol{A}(n-1)\boldsymbol{x}(n-1) + \boldsymbol{\omega}(n) \tag{5.78}$$

where $\boldsymbol{A}(n-1)$ is a time-varying $p \times p$state transition matrix and $\boldsymbol{\omega}(n)$ is a vector of zero-mean white noise processes with

$$E\{\boldsymbol{\omega}(n)\boldsymbol{\omega}^H(n)\} = \begin{cases} \boldsymbol{Q}_\omega(n), k = n \\ \boldsymbol{0}, k \neq n \end{cases}$$

In addition, let $\boldsymbol{y}(n)$ be a vector of observations that are formed according to

$$\boldsymbol{y}(n) = \boldsymbol{C}\boldsymbol{x}(n) + \boldsymbol{v}(n) \tag{5.79}$$

where $\boldsymbol{y}(n)$ is a vector of length q, $\boldsymbol{C}$ is a time-varying $q \times p$ matrix, and $\boldsymbol{v}(n)$ is a vector of zero-mean white noise processes that are statistically independent of $\boldsymbol{\omega}(n)$ with

$$E\{\boldsymbol{v}(n)\boldsymbol{v}^H(n)\} = \begin{cases} \boldsymbol{R}_v(n), k = n \\ \boldsymbol{0}, k \neq n \end{cases}$$

Generalizing the result given in equation (5.57), we expect the optimum linear estimate of $\boldsymbol{x}(n)$ to be expressible in the form

$$\hat{\boldsymbol{x}}(n) = \boldsymbol{A}(n-1)\hat{\boldsymbol{x}}(n-1) + \boldsymbol{K}(n)[\boldsymbol{y}(n) - \boldsymbol{C}\boldsymbol{A}(n-1)\hat{\boldsymbol{x}}(n-1)]$$

With the appropriate Kalman gain matrix $\boldsymbol{K}(n)$, this recursion corresponds to the discrete Kalman filter.

We will now show that the optimum linear recursive estimate of $\boldsymbol{x}(n)$ has this form and derive the optimum Kalman gain $\boldsymbol{K}(n)$ that minimizes the mean-square estimation error. In the following discussion it is assumed that $\boldsymbol{A}(n)$, $\boldsymbol{C}(n)$, $\boldsymbol{Q}_\omega(n)$, and $\boldsymbol{R}_v(n)$ are known.

In our development of the discrete Kalman filter, we will let $\hat{\boldsymbol{x}}(n|n)$ denote the best linear estimate of $\boldsymbol{x}(n)$ at time n given the observations $\boldsymbol{y}(n)$, and we will let $\hat{\boldsymbol{x}}(n|n-1)$ denote the best estimate given the observations up to time $n-1$. The corresponding state estimation errors $\boldsymbol{e}(n|n)$ and $\boldsymbol{e}(n|n-1)$ are as follows

$$\boldsymbol{e}(n|n) = \boldsymbol{x}(n) - \hat{\boldsymbol{x}}(n|n)$$

$$\boldsymbol{e}(n|n-1) = \boldsymbol{x}(n) - \hat{\boldsymbol{x}}(n|n-1)$$

The error covariance matrices, $\boldsymbol{P}(n|n)$, and $\boldsymbol{P}(n|n-1)$ are as follows

$$\begin{aligned} \boldsymbol{P}(n|n) &= E\{\boldsymbol{e}(n|n)\boldsymbol{e}^H(n|n)\} \\ \boldsymbol{P}(n|n-1) &= E\{\boldsymbol{e}(n|n-1)\boldsymbol{e}^H(n|n-1)\} \end{aligned} \tag{5.80}$$

the problem that we would like to solve is the following. Suppose that matrix we are given an estimate $\hat{\boldsymbol{x}}(0|0)$ of the state $\boldsymbol{x}(0)$, and that the error covariance matrix for this estimate, $\boldsymbol{P}(0|0)$ is known. When the measurement $\boldsymbol{y}(1)$ becomes available the goal is to update $\hat{\boldsymbol{x}}(0|0)$ and find the estimate $\hat{\boldsymbol{x}}(1|1)$ of the state at time $n=1$ that minimizes the Mean Square Error

$$\xi(1)=E\{\|\boldsymbol{e}(1|1)\|^2\}=\operatorname{tr}\{\boldsymbol{P}(1|1)\}=\sum_{i=0}^{p-1}E\{|e_i(1|1)|^2\} \tag{5.81}$$

After $\hat{\boldsymbol{x}}(1|1)$ has been determined and the error covariance $\boldsymbol{P}(1|1)$ found, the estimation is repeated for the next observation $\boldsymbol{y}(2)$. Thus, for each $n>0$, given $\hat{\boldsymbol{x}}(n|n-1)$ and $\boldsymbol{P}(n-1|n-1)$, when a new observation, $\boldsymbol{y}(n)$, becomes available, the problem is to find the minimum mean-square estimate $\hat{\boldsymbol{x}}(n|n)$ of the state vector $\boldsymbol{x}(n)$. The solution to this problem will be derived in two steps. First, given $\hat{\boldsymbol{x}}(n-1|n-1)$ we will find $\hat{\boldsymbol{x}}(n|n-1)$, which is the best estimate of $\boldsymbol{x}(n)$ without the observation $\boldsymbol{y}(n)$. Then, given $\boldsymbol{y}(n)$ and $\hat{\boldsymbol{x}}(n|n-1)$ we will estimate $\boldsymbol{x}(n)$ for finding $\hat{\boldsymbol{x}}(n|n)$.

In the first step, since no new measurements are used to estimate $\boldsymbol{x}(n)$, all that is known is that $\boldsymbol{x}(n)$ evolves according to the state equation

$$\boldsymbol{x}(n)=\boldsymbol{A}(n-1)\boldsymbol{x}(n-1)+\boldsymbol{\omega}(n)$$

Since $\boldsymbol{\omega}(n)$ is a zero mean white noise process (and the values of $\boldsymbol{\omega}(n)$ are unknown), then we may predict $\boldsymbol{x}(n)$ as follows

$$\hat{\boldsymbol{x}}(n|n-1)=\boldsymbol{A}(n-1)\hat{\boldsymbol{x}}(n-1|n-1) \tag{5.82}$$

which has an estimation error given by

$$\begin{aligned}\boldsymbol{e}(n|n-1)&=\boldsymbol{x}(n)-\hat{\boldsymbol{x}}(n|n-1)\\&=\boldsymbol{A}(n-1)\boldsymbol{x}(n-1)-\boldsymbol{A}(n-1)\hat{\boldsymbol{x}}(n-1|n-1)\\&=\boldsymbol{A}(n-1)\boldsymbol{e}(n-1|n-1)+\boldsymbol{\omega}(n)\end{aligned} \tag{5.83}$$

and

$$\boldsymbol{P}(n|n-1)=\boldsymbol{A}(n-1)\boldsymbol{P}(n-1|n-1)\boldsymbol{A}^{\mathrm{H}}(n-1)+\boldsymbol{Q}_\omega(n) \tag{5.84}$$

where $\boldsymbol{Q}_\omega(n)$ is the covariance matrix for the noise process $\boldsymbol{\omega}(n)$. This completes the first step of the Kalman filter.

In the second step we incorporate the new measurement $\boldsymbol{y}(n)$ into the estimate $\hat{\boldsymbol{x}}(n|n-1)$. A linear estimate of $\boldsymbol{x}(n)$ that is based on $\hat{\boldsymbol{x}}(n|n-1)$ and $\boldsymbol{y}(n)$ is of the form

$$\hat{\boldsymbol{x}}(n|n)=\boldsymbol{K}'(n)\hat{\boldsymbol{x}}(n|n-1)+\boldsymbol{K}(n)\boldsymbol{y}(n) \tag{5.85}$$

where $\boldsymbol{K}'(n)$ and $\boldsymbol{K}(n)$ are matrices, yet to be specified. The requirement that is imposed on $\hat{\boldsymbol{x}}(n|n)$ is that it be unbiased, $E\{\boldsymbol{e}(n|n)\}=0$, and that it minimizes the Mean Square Error, $E\{\|\boldsymbol{e}(n|n)\|^2\}$. Using equation (5.65) we may express $\boldsymbol{e}(n|n)$ in terms of $\boldsymbol{e}(n|n-1)$ as follows

$$\begin{aligned}\boldsymbol{e}(n|n)&=\boldsymbol{x}(n)-[\boldsymbol{K}'(n)\hat{\boldsymbol{x}}(n|n-1)+\boldsymbol{K}(n)\boldsymbol{y}(n)]\\&=\boldsymbol{x}(n)-\boldsymbol{K}'(n)[\boldsymbol{x}(n)-\boldsymbol{e}(n|n-1)]-\boldsymbol{K}(n)[\boldsymbol{C}(n)\boldsymbol{x}(n)+\boldsymbol{v}(n)]\\&=[\boldsymbol{I}-\boldsymbol{K}'(n)-\boldsymbol{K}(n)\boldsymbol{C}(n)]\boldsymbol{x}(n)+\boldsymbol{K}'(n)\boldsymbol{e}(n|n-1)-\boldsymbol{K}(n)\boldsymbol{v}(n)\end{aligned} \tag{5.86}$$

Since $E\{\boldsymbol{v}(n)\}=0$ and $E\{\boldsymbol{e}(n|n-1)\}=0$, then $\hat{\boldsymbol{x}}(n|n)$ will be unbiased for any $\boldsymbol{x}(n)$ only if the term in brackets is zero

$$K'(n)=I-K(n)C(n)$$

With this constraint, it follows from equation (5.65) that $\hat{x}(n|n)$ has the form

$$\hat{x}(n|n)=[I-K(n)C(n)]\hat{x}(n|n-1)+K(n)y(n)$$

or

$$\hat{x}(n|n)=\hat{x}(n|n-1)+K(n)[y(n)-C(n)\hat{x}(n|n-1)] \tag{5.87}$$

and the error is

$$e(n|n)=[I-K(n)C(n)]e(n|n-1)-K(n)v(n)$$

Thus, the error covariance matrix for $e(n|n)$ is

$$\begin{aligned}P(n|n)&=E\{e(n|n)e^{H}(n|n)\}\\&=[I-K(n)C(n)]P(n|n-1)[I-K(n)C(n)]^{H}+K(n)R_{v}(n)K^{H}(n)\end{aligned} \tag{5.88}$$

Next, we must find the value for the Kalman gain $K(n)$ that minimizes the Mean Square Error

$$\xi(n)=\mathrm{tr}(P(n|n))$$

This may be accomplished in a couple of different ways. Although requiring some special matrix differentiation formulas, we will take the most expedient approach of differentiating $\xi(n)$ with concerning to $K(n)$, setting the derivative to zero, and solving for $K(n)$. Using the matrix differentiation formulas, we have

$$\frac{\mathrm{d}}{\mathrm{d}K}tr(P(n|n))=-2[I-K(n)C(n)]P(n|n-1)C^{H}(n)+2K(n)R_{v}(n)=0$$

Solving for $K(n)$ gives the desired expression for the Kalman gain

$$K(n)=P(n|n-1)C^{H}(n)[C(n)P(n|n-1)C^{H}(n)+R_{v}(n)]^{-1} \tag{5.89}$$

Having found the Kalman gain vector, we may simplify the expression given in equation (5.88) for the error covariance. First, we rewrite the expression for $P(n|n)$ as follows,

$$P(n|n)=E\{e(n|n)e^{H}(n|n)\}=[I-K(n)C(n)]P(n|n-1) \tag{5.90}$$

Thus far we have derived the Kalman filtering equations for recursively estimating the state vector $x(n)$. All that needs to be done to complete the recursion is to determine how the recursion should be initialized at time $n=0$. Since the value of the initial state is unknown, in the absence of any observed data at time $n=0$, the initial estimate is chosen to be

$$\hat{x}(0|0)=E\{x(0)\}$$

and, for the initial value for the error covariance matrix, we have

$$P(0|0)=E\{x(0)x^{H}(0)\}$$

This choice for the initial conditions makes $\hat{x}(0|0)$ an unbiased estimate of $x(0)$ and ensures that $\hat{x}(n|n)$ will be unbiased for all n (recall that the Kalman filtering update equations were derived with the constraint that $\hat{x}(n|n)$ is unbiased). This completes the derivation of the discrete Kalman filter which is summarized in Table 5.4. One interesting property to note about the Kalman filter is that the Kalman gain $K(n)$ and the error covariance matrix $P(n|n)$ do not depend on the data $x(n)$. Therefore, it is possible for both of these terms to be computed off-line prior to any filtering.

Table 5.4 The Discrete Kalman Filter

State Equation
$$\mathbf{x}(n) = \mathbf{A}(n-1)\mathbf{x}(n-1) + \boldsymbol{\omega}(n)$$
Observation Equation
$$\mathbf{y}(n) = \mathbf{C}\mathbf{x}(n) + \mathbf{v}(n)$$
Initialization
$$\hat{\mathbf{x}}(0\|0) = E\{\mathbf{x}(0)\}$$ $$\mathbf{P}(0\|0) = E\{\mathbf{x}(0)\mathbf{x}^{\mathrm{H}}(0)\}$$
Computation
1. Prediction
$$\hat{\mathbf{x}}(n\|n-1) = \mathbf{A}(n-1)\hat{\mathbf{x}}(n-1\|n-1)$$ $$\mathbf{P}(n\|n-1) = \mathbf{A}(n-1)\mathbf{P}(n-1\|n-1)\mathbf{A}^{\mathrm{H}}(n-1) + \mathbf{Q}_{\omega}(n)$$
2. Measurement update
$$\mathbf{K}(n) = \mathbf{P}(n\|n-1)\mathbf{C}^{\mathrm{H}}(n)\left[\mathbf{C}(n)\mathbf{P}(n\|n-1)\mathbf{C}^{\mathrm{H}}(n) + \mathbf{R}_{v}(n)\right]^{-1}$$ $$\hat{\mathbf{x}}(n\|n) = \hat{\mathbf{x}}(n\|n-1) + \mathbf{K}(n)\left[\mathbf{y}(n) - \mathbf{C}(n)\hat{\mathbf{x}}(n\|n-1)\right]$$ $$\mathbf{P}(n\|n) = \left[\mathbf{I} - \mathbf{K}(n)\mathbf{C}(n)\right]\mathbf{P}(n\|n-1)$$

5.6 Summary

In this chapter, we considered the problem of designing the optimum filter. The first problem is to design an optimal FIR filter that minimizes the mean square estimation error $\xi = E\{|d(n)-\hat{d}(n)|\}^2$. Assuming that $d(n)$ and $x(n)$ are jointly wide-sense stationary processes with known autocorrelation $r_x(k)$ and cross-correlation $r_{dx}(k)$, the solution is given by the Wiener-Hopf equations, which are a set of linear Toeplitz equations.

Since the Wiener-Hopf equations apply to the estimation of any process $d(n)$, we then considered the special cases of filtering, linear prediction, and noise cancellation. Next, we described the design of an IIR Wiener filter. In the absence of a causality constraint imposed on the solution, we find that these filters are generally noncausal and, therefore unrealizable. As a result, these filters are generally unsuitable for real-time signal processing applications.

However, assuming that the power spectrum of $x(n)$ and the cross-power spectral density between $x(n)$ and $d(n)$ are known, this can be used to set an upper bound on the performance of an FIR Wiener filter or a causal IIR Wiener filter. We introduced the design of a causal IIR Wiener filter. We find that, by imposing a causality constraint on the filter, it becomes necessary to perform a spectral factorization of the power spectrum of the input process $x(n)$. Therefore, the design of these filters is usually much more difficult compared to the design of FIR Wiener filters or non-causal Wiener filters.

In addition, we also explored how to design filters based on the desired signal and observations when the a priori statistical properties are unknown. Relying on the LSE criterion, we implemented FIR filters based on observations.

Finally, we briefly discussed the problem of recursive filtering and derived the discrete Kalman filter. Unlike the Wiener filter, the Kalman filter may be used for nonstationary processes as well as stationary ones, and may be initialized to begin operating at time $n = 0$.

Exercises

1. Next we present an example of Kalman filtering. First assume that there is a random path with a temporal order relationship in 3D space with a length of 5,000 steps. The noise with variance 2 is superimposed on the path, i. e., the observed noise $v(n)$ has $R = 2$. Further, the process error covariance of Kalman filter $Q = 0.001$, and the general setting of the experiments is as follows.

Simulation:

Total steps = 5,000;

Movement path: Random;

State transition matrix: A=eye(6)+diag(ones(1,3),3);

Measurement matrix: H=eyes(3,6);

Process error covariance Q=diag(ones(6,1) *0.001)

Measurement error covariance R=diag(ones(3,1) *2)

Solution:

Based on the above settings, the simulation is carried out using Matlab, here is the relevant Matlab source code:

Updating:

for k=1:Total steps

```
for k=1:Total steps
    X_=kf_params.A * kf_params.X+kf_params.B * kf_params.U;
    P_=kf_params.A * kf_params.P * kf_params.A'+kf_params.Q;
    kf_params.K=P_ * kf_params.H' / (kf_params.H * P_ * kf_params.H'+kf_params.R);
    kf_params.X=X_+kf_params.K * (kf_params.Z-kf_params.H * X);
    kf_params.P=P_-kf_params.K * kf_params.H * P_'
end
```

2. In this example we consider a "desired" signal $s(n)$, whose generation process is performed through the process $s(n) = -0.8w(n-1) + w(n)$, where $w(n)$ is white noise and follows the distribution of $N(0,\sigma_w^2)$, where $\sigma_w^2 = 0.3$. This signal is passed through the causal system $H(z) = 1 - 0.9z^{-1}$ whose output $y(n)$ is corrupted by additive white noise $v(n) \sim WN(0,\sigma_v^2)$, and $\sigma_v^2 = 0.1$. Note that the processes $w(n)$ and $v(n)$ should be uncorrelated.

Question:

Design a second-order optimum FIR filter that estimates $s(n)$ from the signal $x(n) = y(n) + v(n)$ and determine $\boldsymbol{c}_0$ and P_0

Solution:

Relevant Matlab source code:

Paramas Initial

```
Var_w=0.3;
Var_v=0.1;
H1=[1,-0.8];
lengthH1=length(H1);
H=[1,-0.9];
```

Computation

```
rh1=conv(h1,fliplr(h1)); lrh1=length(rh1);
rs=var_w * rh1; rs=rs(lh1:end); Ps=rs(1);
h2=conv(h,h1); lh2=length(h2);
rx=var_w * conv(h2,fliplr(h2))+var_v * [zeros(1,lh2-1),1,zeros(1,lh2-1)];
rx=rx(lh2:end);
lrh1c=round((lrh1-1)/2)+1;
rxs=var_w * conv(h,rh1); rxs=rxs(lrh1c:end);
```

Second-order optimal FIR filter design

```
M=2;
Rx=toeplitz(rx(1:M),rx(1:M));
dxs=rxs(1:M)';
co=Rx \dxs;
```

Output

```
Second-order (i.e., length M=2) optimal FIR filter design
Optimal FIR filter coefficients: 0.3255,-0.2793
Optimal error=0.0709
```

References

[1] R. G. Brown, *Introduction to Random Signal Analysis and Kalman Filtering*, John Wiley & Sons, New York, 1983.

[2] T. Kailath, "An innovations approach to least-squares estimation–Part I: Linear filtering in additive noise," *IEEE Trans. Autom. Control*, vol. AC–13, pp. 641–655, 1968.

[3] R. E. Kalman, "A new approach to linear filtering and prediction problems", *Trans. ASME*, *J. Basic Eng.* , Ser. 820, pp. 35–45, March 1960.

[4] H. W. Sorensen, "Least-squares estimation: From Gauss to Kalman," *IEEE Spectrum*, vol. 7, pp. 63–68, July 1970.

[5] H. W. Sorensen, ed. , *Kalman Filtering: Theory and Application*, IEEE Press, New York, 1985.

Chapter 6

Adaptive Filters

6.1 Introduction

In previous chapters, we have considered a variety of different problems including signal modeling, parametric spectrum estimation, and optimal filtering. In each case, we have made an important assumption: the signal analyzed is stationary. For example, we assume that the signal to be modeled can be approximated as the output of a linear time-invariant (LTI) filter, whose input is either a unit impulse in deterministic signal modeling or stationary white noise in random signal modeling. In Chapter 5, we discussed the problem of designing a linear shift-invariant filter solving the minimum mean square estimate of a wide-sense stationary process. Unfortunately, the methods and techniques considered so far become inappropriate due to the fact that the signals we use in real situations are nonstationary. One way of dealing with this problem is to process these nonstationary signals over intervals, where the signals can be assumed as a stationary process. However, this approach is limited for several reasons. First, for rapidly changing processes, the time interval over which the process is assumed to be a steady state may be too small to provide sufficient precision or resolution to estimate the interested parameters. Second, this approach does not easily accommodate step changes within the analysis interval. Third, and perhaps most importantly, this solution imposes an incorrect model on the data, i.e., piecewise stationary. Therefore, a better approach is to make non-stationary assumptions at the outset.

First, let us reconsider the problem of Wiener filtering in the context of a nonstationary process. Specifically, let $w(n)$ denotes the unit impulse response of the FIR Wiener filter that produces minimum mean-square estimate of a desired process $d(n)$, which can be expressed as

$$\hat{d}(n) = \sum_{k=0}^{k=p} w(k)x(n-k) \tag{6.1}$$

As we can know that if $x(n)$ and $d(n)$ are jointly wide-sense stationary processes, with $e(n) = d(n) - \hat{d}(n)$, then the filter optimal coefficients used by minimizing the Mean Square Error $E\{|e_{(n)}|^2\}$ are found by solving the Wiener-Hopf equations

$$\boldsymbol{R}_x \boldsymbol{w} = \boldsymbol{r}_{dx} \tag{6.2}$$

However, if $d(n)$ and $x(n)$ are nonstationary, then the filter coefficients that minimize $E\{|e_{(n)}|^2\}$ will depend on n, and the filter will be shift-varying, i. e.,

$$\hat{d}(n) = \sum_{k=0}^{k=p} w_n(k)x(n-k) \tag{6.3}$$

where $w_n(k)$ is the value of the k th filter coefficient at time n. Using vector expression, this estimate may be expressed as

$$\hat{d}(n) = w_n^{\mathrm{T}} x(n) \tag{6.4}$$

where

$$\boldsymbol{w}_{\mathrm{n}} = [w_n(0), w_n(1), \cdots, w_n(p)]^{\mathrm{T}}$$

is the vector of filter coefficients at time n, and the input vector is

$$\boldsymbol{x}_{\mathrm{n}} = [x_n(n), x_n(n-1), \cdots, x_n(n-p)]^{\mathrm{T}}$$

In some respects, the design of a shift-varying (adaptive) filter is much more difficult than the design of a (shift-invariant) Wiener filter. This is because, for each value of n, it is necessary to find the set of optimal filter coefficients $w_n(k)$, for $k = 0, 1, 2, \cdots, p$. However, the problem may be simplified considerably if we relax the requirement that $\boldsymbol{w}_n$ minimizes the Mean Square Error at each time n and instead consider a coefficient update equation of the form

$$\boldsymbol{w}_{n+1} = \boldsymbol{w}_n + \Delta\boldsymbol{w}_n \tag{6.5}$$

where $\Delta\boldsymbol{w}_n$ is the correction applied to the filter coefficients $\boldsymbol{w}_n$ at time n to form a new set of coefficients $\boldsymbol{w}_{n+1}$ at time $n+1$. The heart of the adaptive filters that we will design in this chapter is this update equation. The design of an adaptive filter consists of defining how this correction is to be formed. Even for stationary processes, there are several reasons why we choose to implement a time-invariant Wiener filter using equation (6.5). First, if the filter has a large order p, then solving the Wiener-Hopf equations directly may be difficult and impractical. Second, if $\boldsymbol{R}_x$ is ill-conditioned (almost singular), then the solution to Wiener-Hopf equations will be numerically sensitive to round-off errors and finite precision effects. Finally, the most important reason is that solution of the Wiener-Hopf equations requires the knowledge of the autocorrelation $r_x(k)$ and the cross-correlation $r_{dx}(k)$. Since these statistical ensemble averages are often unknown, then it is necessary to estimate them from measurements of the processes. Although we can use estimates such as

$$\hat{r}_x(k) = \frac{1}{N}\sum_{n=0}^{N-1} x(n)x^*(n-k)$$

$$\hat{r}_{dx}(k) = \frac{1}{N}\sum_{n=0}^{N-1} d(n)x^*(n-k) \tag{6.6}$$

doing so would result in a delay of N samples. More importantly, in an environment where the ensemble averages change in time, these estimates would need to be constantly updated.

The key to an adaptive filter is the set of guidelines, or algorithms, that define how the correction $\Delta\boldsymbol{w}_n$ should be formed. Although it is not yet clear what this correction should be, it is clear that the sequence of corrections should reduce the Mean Square Error. In fact, whatever algorithm is used, the adaptive filter should have the following characteristics:

1. In a stationary environment, the adaptive filter should produce a sequence of corrections $\Delta\boldsymbol{w}_n$ in such a way that the coefficients $\boldsymbol{w}_n$ converge to the solution to the Wiener-Hopf equations

$$\lim_{n\to\infty} \boldsymbol{w}_n = \boldsymbol{R}_x^{-1}\boldsymbol{r}_{dx}$$

2. It should not be necessary to know the statistical characteristics $r_x(k)$ and $r_{dx}(k)$ in order to compute $\Delta \mathbf{w}_n$. The estimation of these statistics should be "built into" the adaptive filter.

3. For nonstationary signals, the filter should be able to adapt to the changing statistics and "track" the solution as it evolves in time.

An important issue in the implementation of adaptive filters is that the error signal $e(n)$ is available to the adaptive algorithm. This is because $e(n)$ allows the filter to measure its performance and gives guidance on how to modify the filter coefficients. In some applications, the acquisition of $d(n)$ is straightforward, which makes the evaluation of $e(n)$ easy. For example, consider the problem of system identification shown in Figure 6. 1 in which a system produces an output $d(n)$ in response to a known input $x(n)$. The goal is to develop a model $W_n(z)$ for the system, $W_n(z)$ producing a response $d(n)$ that is as close as possible to $d(n)$. To consider the observation noise in the measurement of the system output, an additive noise source $v(n)$ is shown in the figure.

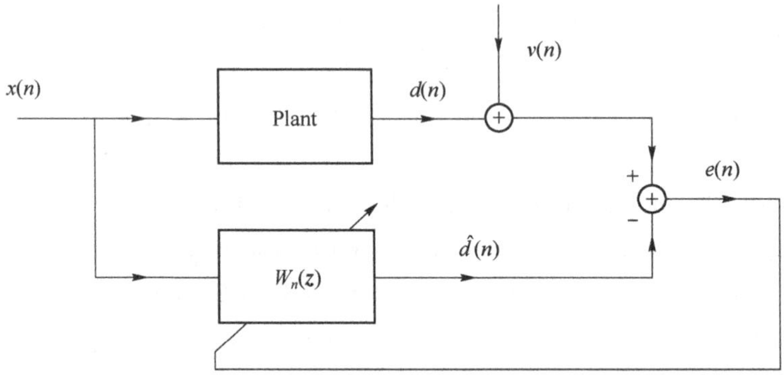

Figure 6. 1 adaptive filtering application in system identification

In this chapter, we will consider a variety of different methods for designing and implementing adaptive filters. As we will see, the efficiency of the adaptive filter and its performance in estimating $d(n)$ will depend on a number of factors including the type of filters (FIR or IIR), and the way in which the performance measure is defined (Mean Square Error, Least Squares error). The organization of this chapter is as follows. In Section 6. 2, we start with the LMS adaptive filters that are based on the method of the steepest descent, in Section 6. 3, the Affine Projection Adaptive filter (APA) algorithm is also introduced, and in Section 6.4, the Recursive Least Squares(RLS) algorithm is developed.

6. 2 LMS Adaptive Filter

6. 2. 1 The Steepest Descent Adaptive Filter

When designing an FIR adaptive filter, the goal is to find the vector $\mathbf{w}_n$ at time n that minimizes the quadratic function

$$\xi(n) = E\{|e_{(n)}|^2\}$$

Although the vector that minimizes $\xi(n)$ may be found by setting the derivatives of $\xi(n)$ with respect to $w^*(k)$ equal to zero, another approach is to search for the solution using the method of steepest descent. The method of steepest descent is an iterative procedure that has been used to find extrema of nonlinear functions since before the time of Newton. The basic idea of this method is as follows. Let $\boldsymbol{w}_n$ be an estimate of the vector that minimizes the Mean Square Error $\xi(n)$ at time n. At time $n+1$, a new estimate is formed by adding a correction to $\boldsymbol{w}_n$ that is designed to bring $\boldsymbol{w}_n$ closer to the desired solution. The correction involves taking a step of size μ in the direction of maximum descent down the quadratic error surface. Since the gradient vector points in the direction of steepest ascent, and the direction of steepest descent points in the negative gradient direction. Thus, the update equation for $\boldsymbol{w}_n$ is

$$\boldsymbol{w}_{n+1} = \boldsymbol{w}_n - \mu \nabla\xi(n)$$

The step size μ affects the rate at which the weight vector moves down the quadratic surface and must be a positive number (a negative value for μ would move the weight vector up the quadratic surface in the direction of maximum ascent and would result in an increase in the error). For very small values of μ, the correction to $\boldsymbol{w}_n$ is small and the movement down the quadratic surface is slow. As μ is increased, the rate of descent increases. However, there is an upper limit on how large the step size may be. For values of μ that exceed this limit, the trajectory of $\boldsymbol{w}_n$ becomes unstable and unbounded. The steepest descent algorithm may be summarized as follows:

1. Initialize the steepest descent algorithm with an initial estimate $\boldsymbol{w}_0$, of the optimum weight vector $\boldsymbol{w}$.

2. Evaluate the gradient of $\xi(n)$ at the current estimate $\boldsymbol{w}_n$, of the optimum weight vector.

3. Update the estimate at time n by adding a correction that is formed by taking a step of size μ in the negative gradient direction

$$\boldsymbol{w}_{n+1} = \boldsymbol{w}_n - \mu \nabla\xi(n).$$

4. Go back to (2) and repeat the process.

Let us now evaluate the gradient vector $\nabla\xi(n)$. Assuming that $\boldsymbol{w}$ is complex, the gradient is the derivative of $E\{|e_{(n)}|^2\}$ with respect to $\boldsymbol{w}^*$. With

$$\nabla\xi(n) = \nabla E\{|e_{(n)}|^2\} = E\{\nabla|e_{(n)}|^2\} = E\{e(n)\nabla e^*(n)\}$$

and

$$\nabla e^*(n) = -\boldsymbol{x}^*(n)$$

it follows that

$$\nabla\xi(n) = -E\{e(n)\boldsymbol{x}^*(n)\}$$

Thus, with a step size of μ, the steepest descent algorithm becomes

$$\boldsymbol{w}_{n+1} = \boldsymbol{w}_n + \mu E\{e(n)\boldsymbol{x}^*(n)\} \quad (6.7)$$

To see how this steepest descent update equation for $\boldsymbol{w}_n$ performs, let us consider what happens in the case of stationary processes. If $x(n)$ and $d(n)$ are jointly WSS then

$$\begin{aligned} E\{e(n)\boldsymbol{x}^*(n)\} &= E\{d(n)\boldsymbol{x}^*(n)\} - E\{\boldsymbol{w}_n^{\mathrm{T}}\boldsymbol{x}(n)\boldsymbol{x}^*(n)\} \\ &= \boldsymbol{r}_{dx} - \boldsymbol{R}_x\boldsymbol{w}_n \end{aligned}$$

and the steepest descent algorithm becomes

$$\boldsymbol{w}_{n+1} = \boldsymbol{w}_n + \mu(\boldsymbol{r}_{dx} - \boldsymbol{R}_x \boldsymbol{w}_n) \tag{6.8}$$

Note that if $\boldsymbol{w}_n$ is the solution to the Wiener-Hopf equations, $\boldsymbol{w}_n = \boldsymbol{R}_x^{-1}\mathbf{r}_{dx}$, then thecorrection term is zero and $\boldsymbol{w}_{n+1} = \boldsymbol{w}_n$ for all n. Of greater interest, however, is how the weights evolve in time, beginning with an arbitrary initial weight vector $\boldsymbol{w}_0$. The following property defines what is required for to converge $\mathbf{w}_n$ to $\mathbf{w}$.

Property 1: For jointly wide-sense stationary processes, $d(n)$ and $x(n)$, the steepest descentadaptive filter converges to the solution to the Wiener-Hopf equations

$$\lim_{n\to\infty} \boldsymbol{w}_n = \boldsymbol{R}_x^{-1}\boldsymbol{r}_{dx}$$

if the step size satisfies the condition

$$0 < \mu < \frac{2}{\lambda_{\max}} \tag{6.9}$$

where $\lambda_{\max}$ is the maximum eigenvalue of the autocorrelation matrix $\boldsymbol{R}_x$.

To establish this property, we begin by rewriting equation (6.8) as follows

$$\boldsymbol{w}_{n+1} = (\boldsymbol{I} - \mu\boldsymbol{R}_x)\boldsymbol{w}_n + \mu\boldsymbol{r}_{dx} \tag{6.10}$$

Subtracting $\boldsymbol{w}$ from both sides of this equation and using the fact that $\boldsymbol{r}_{dx} = \boldsymbol{R}_x\boldsymbol{w}$ we have

$$\boldsymbol{w}_{n+1} = (\boldsymbol{I} - \mu\boldsymbol{R}_x)\boldsymbol{w}_n + \mu\boldsymbol{R}_x\boldsymbol{w} - \boldsymbol{w} = [\boldsymbol{I} - \mu\boldsymbol{R}_x](\boldsymbol{w}_n - \boldsymbol{w}) \tag{6.11}$$

If we let $\boldsymbol{c}_n$ be the weight error vector,

$$\boldsymbol{c}_n = \boldsymbol{w}_n - \boldsymbol{w} \tag{6.12}$$

then equation (6.11) becomes

$$\boldsymbol{c}_{n+1} = [\boldsymbol{I} - \mu\boldsymbol{R}_x]\boldsymbol{c}_n \tag{6.13}$$

Note that, unless $\boldsymbol{R}_x$ is a diagonal matrix, there will be cross-coupling between the coefficients of the weight error vector. However, we may decouple these coefficients by diagonalizing the autocorrelation matrix as follows. Using the spectral theorem, we can factor the autocorrelation matrix as

$$\boldsymbol{R}_x = \boldsymbol{V}\boldsymbol{\Lambda}\boldsymbol{V}^{\mathrm{H}}$$

where $\boldsymbol{\Lambda}$ is a diagonal matrix containing the eigenvalues of $\boldsymbol{R}_x$, and $\boldsymbol{V}$ is a matrix whose columns are the eigenvectors of $\boldsymbol{R}_x$. Since $\boldsymbol{R}_x$ is Hermitian and non-negative definition, the eigenvalues are real and non-negative, $\lambda_k \geqslant 0$, and the eigenvectors may be chosen to be orthonormal, $\boldsymbol{V}\boldsymbol{V}^{\mathrm{H}} = \boldsymbol{I}$, i. e., $\boldsymbol{V}$ is unitary. Incorporating this factorization into equation (6.13) leads to

$$\boldsymbol{c}_{n+1} = (\boldsymbol{I} - \mu\boldsymbol{V}\boldsymbol{\Lambda}\boldsymbol{V}^{\mathrm{H}})\boldsymbol{c}_n$$

Using the unitary property of $\boldsymbol{V}$, we have

$$\boldsymbol{c}_{n+1} = (\boldsymbol{V}\boldsymbol{V}^{\mathrm{H}} - \mu\boldsymbol{V}\boldsymbol{\Lambda}\boldsymbol{V}^{\mathrm{H}})\boldsymbol{c}_n = \boldsymbol{V}(\boldsymbol{I} - \mu\boldsymbol{\Lambda})\boldsymbol{V}^{\mathrm{H}}\boldsymbol{c}_n$$

Multiplying $\boldsymbol{V}^{\mathrm{H}}$ for both sides of the equation gives

$$\boldsymbol{V}^{\mathrm{H}}\boldsymbol{c}_{n+1} = (\boldsymbol{I} - \mu\boldsymbol{\Lambda})\boldsymbol{V}^{\mathrm{H}}\boldsymbol{c}_n \tag{6.14}$$

If we define

$$\boldsymbol{u}_n = \boldsymbol{V}^{\mathrm{H}}\boldsymbol{c}_n \tag{6.15}$$

then equation (6.14) becomes

$$\boldsymbol{u}_{n+1} = (\boldsymbol{I} - \mu\boldsymbol{\Lambda})\boldsymbol{u}_n$$

Equation (6.15) represents a rotation of the coordinate system for the weight error vector $\boldsymbol{c}_n$ with the new axes aligned with the eigenvectors $\boldsymbol{v}_k$ of the autocorrelation matrix. With an initial weight vector $\boldsymbol{u}_0$, it follows that

$$\boldsymbol{u}_n = (\boldsymbol{I}-\mu\boldsymbol{\Lambda})^n \boldsymbol{u}_0 \tag{6.16}$$

Since $(\boldsymbol{I}-\mu\boldsymbol{\Lambda})$ is a diagonal matrix, then the k th component of $\boldsymbol{u}_n$ may be expressed as

$$u_n(k) = (1 - \mu\lambda_k)^n u_0(k) \tag{6.17}$$

In order for $\boldsymbol{w}_n$ to converge to $\mathbf{w}$ it is necessary that the weight error vector $\boldsymbol{c}_n$ converge to zero and, therefore, that $\boldsymbol{u}_n = \boldsymbol{V}^{\mathrm{H}}\boldsymbol{c}_n$ converge to zero. This will occur for any $\boldsymbol{u}_0$ if and only if

$$|1 - \mu\lambda_k| < 1 \text{ for } k = 0,1,2,\cdots,p$$

which places the following restriction on the step size μ

$$0 < \mu < \frac{2}{\lambda_{\max}}$$

as was to be shown.

Although for stationary processes the steepest descent adaptive filter converges to the solution to the Wiener-Hopf equations when $\mu < \frac{2}{\lambda_{\max}}$, this algorithm is mainly of theoretical interest and has little use in adaptive filtering. The reason for this is that it is necessary to have the prior knowledge of $E\{e(n)\boldsymbol{x}^*(n)\}$ in order to do the computation. For stationary processes this requires that the autocorrelation matrix of $x(n)$, and the cross-correlation between $d(n)$ and $x(n)$ should be known. In most applications, these ensemble averages are unknown and must be estimated from the data. In the next section, we consider the LMS algorithm, which incorporates an estimate of the expectation $E\{e(n)\boldsymbol{x}^*(n)\}$ into the adaptive algorithm.

6.2.2 The LMS Algorithms

In the previous section, we developed the steepest descent adaptive filter, which has a weight-vector update equation given by

$$\boldsymbol{w}_{n+1} = \boldsymbol{w}_n + \mu E\{e(n)\boldsymbol{x}^*(n)\} \tag{6.18}$$

A practical limitation with this algorithm is that the expectation $E\{e(n)\boldsymbol{x}^*(n)\}$ is generally unknown. Therefore, it must be replaced with an estimate such as the sample mean

$$\hat{E}\{e(n)\boldsymbol{x}^*(n)\} = \frac{1}{L}\sum_{l=0}^{L-1} e(n-l)\boldsymbol{x}^*(n-l) \tag{6.19}$$

Incorporating this estimate into the steepest descent algorithm, the update for $\boldsymbol{w}_n$ becomes

$$\boldsymbol{w}_{n+1} = \boldsymbol{w}_n + \frac{\mu}{L}\sum_{l=0}^{L-1} e(n-l)\boldsymbol{x}^*(n-l) \tag{6.20}$$

A special case of equation (6.20) occurs if we use a one-point sample mean $L = 1$

$$\hat{E}\{e(n)\boldsymbol{x}^*(n)\} = e(n)\boldsymbol{x}^*(n) \tag{6.21}$$

In this case, the weight vector update will become

$$\boldsymbol{w}_{n+1} = \boldsymbol{w}_n + \mu e(n)\boldsymbol{x}^*(n) \tag{6.22}$$

and is known as the LMS algorithm. The simplicity of the algorithm comes from the fact that the update for the k th coefficient,

$$w_{n+1}(k) = w_n(k) + \mu e(n)x^*(n-k)$$

And we can summarize the LMS algorithm as following Table 6.1. The LMS algorithm requires only one multiplication and one addition (the value for $\mu e(n)$ needs only to be computed once and may be used for all of the coefficients). Therefore, an LMS adaptive filter with $p+1$ coefficients requires $p+1$ multiplications and $p+1$ additions to update the filter coefficients. In addition, an addition is necessary to compute the error $e(n) = d(n) - y(n)$ and one multiplication is needed to form the product $\mu e(n)$. Finally, $p+1$ multiplications and p additions are necessary to calculate the output $y(n)$ of the adaptive filter. Thus, a total of $2p+3$multiplications and $2p+2$additions per output point are required.

Table 6.1 The LMS Algorithms for a p-th order FIR adaptive filter

Parameters: p = Filter order μ = Step size Initialization: $\boldsymbol{w}_0 = 0$ Computation: for $n = 0, 1, 2, \cdots$ $y = \boldsymbol{w}_n^{\mathrm{T}} \boldsymbol{x}(n)$ $e(n) = d(n) - y(n)$ $\boldsymbol{w}_{n+1} = \boldsymbol{w}_n + \mu e(n) \boldsymbol{x}(n)$

6.2.3 Convergence of the LMS Algorithms

In estimating the ensemble average $\hat{E}\{e(n)\boldsymbol{x}^*(n)\}$ with a one-point sample average $e(n)\boldsymbol{x}^*(n)$, the LMS algorithm replaces the gradient in the steepest descent algorithm,

$$\nabla\xi(n) = -E\{e(n)\boldsymbol{x}^*(n)\}$$

with an estimated gradient

$$\nabla\xi(n) = -e(n)\boldsymbol{x}^*(n) \tag{6.23}$$

When this is done, the correction that is applied to $\boldsymbol{w}_n$ is generally not aligned with the direction of steepest descent. However, since the gradient estimate is unbiased

$$E\{\hat{\nabla}\xi(n)\} = -E\{e(n)x^*(n)\} = \nabla\xi(n)$$

then the correction applied on the average will be in the direction of the steepest descent. Since $\boldsymbol{w}_n$ is a vector of random variables, the convergence of the LMS algorithm must be considered within a statistical framework. Therefore, we assume that $\boldsymbol{x}(n)$ and $d(n)$ are jointly wide-sense stationary processes, and will determine when the coefficients $\boldsymbol{w}_n$ converge in the mean to $\boldsymbol{w} = \boldsymbol{R}_x^{-1}\boldsymbol{r}_{dx}$, i.e.,

$$\lim_{n\to\infty} E\{\boldsymbol{w}_n\} = \boldsymbol{w} = \boldsymbol{R}_x^{-1}\boldsymbol{r}_{dx}$$

We substitute equation (6.3) and equation (6.4) into the LMS coefficient update equation as follows

$$\boldsymbol{w}_{n+1} = \boldsymbol{w}_n + \mu[d(n) - \boldsymbol{w}_n^{\mathrm{T}}\boldsymbol{x}(n)]\boldsymbol{x}^*(n)$$

Taking the expected value, we have

$$E\{\boldsymbol{w}_{n+1}\} = E\{\boldsymbol{w}_n\} + \mu \mathrm{E}\{d(n)\boldsymbol{x}^*(n)\} - \mu E\{\boldsymbol{x}^*(n)\boldsymbol{x}^{\mathrm{T}}(n)\boldsymbol{w}_n\} \tag{6.24}$$

Although the last term in equation (6.24) is not easy to evaluate, it may be simplified considerably if we make the following independence assumption:

Independence Assumption. The data $\boldsymbol{x}(n)$ and the LMS weight vector $\boldsymbol{w}_n$ are statistically independent. With this assumption, equation (6.24) becomes

$$\begin{aligned} E\{\boldsymbol{w}_{n+1}\} &= E\{\boldsymbol{w}_n\} + \mu E\{d(n)\boldsymbol{x}^*(n)\} - \mu E\{\boldsymbol{x}^*(n)\boldsymbol{x}^{\mathrm{T}}(n)\}E\{\boldsymbol{w}_n\} \\ &= (\boldsymbol{I} - \mu \boldsymbol{R}_x)E\{\boldsymbol{w}_n\} + \mu \boldsymbol{r}_{dx} \end{aligned} \tag{6.25}$$

which is the same as equation (6.10) for the weight vector in the steepest descent algorithm. Therefore, the analysis for the steepest descent algorithm is applicable to $E\{\boldsymbol{w}_{n+1}\}$. In particular, it follows from equation (6.16) that

$$E\{\boldsymbol{u}_n\} = (\boldsymbol{I} - \mu\boldsymbol{\Lambda})^n \boldsymbol{u}_0 \tag{6.26}$$

where

$$\boldsymbol{u}_n = \boldsymbol{v}^{\mathrm{H}}[\boldsymbol{w}_n - \boldsymbol{w}]$$

Since $\boldsymbol{w}_n$ will converge in the mean to $\boldsymbol{w}$ if $E\{\boldsymbol{u}_n\}$ converges to zero, then we have the following property:

Property 2. For jointly wide-sense stationary processes, the LMS algorithm converges in the mean if

$$0<\mu<\frac{2}{\lambda_{\max}}$$

and the independence assumption is satisfied.

Although the property 2 places a bound on the step size for convergence in the mean, this bound is of limited use for two reasons. First, it is generally acknowledged that the upper bound is too large to ensure stability of the LMS algorithm since it is not sufficient to guarantee that the coefficient vector will remain bounded for all n. For example, although this bound ensures that $E\{\boldsymbol{w}_n\}$ converges, it places no constraints on how large the variance of $\boldsymbol{w}_n$ may become. Second, since the upper bound is expressed in terms of the largest eigenvalue of $\boldsymbol{R}_x$, using this bound requires $\boldsymbol{R}_x$ to be known. If this matrix is unknown, then it is necessary to estimate $\lambda_{\max}$. One way around this difficulty is to use the fact that $\lambda_{\max}$ may be upper bounded by the trace of $\boldsymbol{R}_x$,

$$\lambda_{\max} \leqslant \sum_{k=0}^{p} \lambda_k = tr(\boldsymbol{R}_x)$$

Therefore, if $x(n)$ is wide-sense stationary, then $\boldsymbol{R}_x$ is a Toeplitz matrix and the trace of the matrix becomes

$$tr(\boldsymbol{R}_x) = (p+1)r_x(0) = (p+1)E\{|x(n)|^2\}$$

As a result, the bound of μ may be replaced with the more conservative bound

$$0 < \mu < \frac{2}{(p+1)E\{|x(n)|^2\}} \tag{6.27}$$

Although we have simply replaced one unknown with another, $E\{|x(n)|^2\}$ is more easily estimated since it represents the power in $x(n)$. For example, $E\{|x(n)|^2\}$ could be estimated by using an average such as

$$E\{|x(n)|^2\} = \frac{1}{N}\sum_{k=0}^{N-1} |x(n-k)|^2$$

Once meeting with the above properties, as the weight vector begins to converge in the mean, the coefficients begin to fluctuate about their optimum values.

6.2.4 LMS Misadjustment (EMSE)

As we saw in the previous section, when the weight vector starts to converge in the mean, the coefficients begin to fluctuate around their optimal values. These fluctuations are caused by the noise gradient vector used to form the corrections to $\boldsymbol{w}_n$. The result is that the variance of the weight error vector does not go to zero and the mean square error is larger than the minimum mean square error, which is referred to as Excess Mean Squared Error (EMSE). Specifically, when $\boldsymbol{w}_n$ oscillates about $\boldsymbol{w}=\boldsymbol{R}_x^{-1}\boldsymbol{r}_{dx}$, the corresponding mean square error has a value, that exceeds the minimum mean squared error on the average. To quantify this Excess Mean Square Error, we will write the error at time n as follows:

$$e(n)=d(n)-\boldsymbol{w}_n^{\mathrm{T}}\boldsymbol{x}(n)=d(n)-(\boldsymbol{w}+\boldsymbol{c}_n)^{\mathrm{T}}\boldsymbol{x}(n)=e_{\min}(n)-\boldsymbol{c}_n^{\mathrm{T}}\boldsymbol{x}(n)$$

where $e_{\min}(n)$ is the error derived in the optimal filtering case, i. e.

$$e_{\min}(n)=d(n)-\boldsymbol{w}^{\mathrm{T}}\boldsymbol{x}(n)$$

Assuming that the adaptive filter is in steady state, i. e., $E\{\boldsymbol{c}_n\}=0$, the mean square error can be expressed as

$$\xi(n)=E\{|e_n|^2\}=\xi_{\min}+\xi_{ex}(n)$$

in which

$$\xi_{\min}=E\{|e_{\min}(n)|^2\} \tag{6.28}$$

is the minimum mean squared error, and $\xi_{ex}(n)$ is the Excess Mean Squared Error, which depends on $\boldsymbol{x}(n)$, $\boldsymbol{c}_n e(n)$ and $d(n)$. Although, $\xi_{ex}(n)$ is not easy to calculate, the following property can be derived by the independence assumption.

Property 3. The mean squared error $\zeta(n)$ $\zeta(n)$ converges to a steady state value of

$$\xi(\infty)=\xi_{\min}+\xi_{ex}(\infty)=\xi_{\min}\frac{1}{1-\mu\sum_{k=0}^{p}\frac{\lambda_k}{2-\mu\lambda_k}}$$

At this point, the LMS algorithm converges in the mean square if and only if the step size μ satisfies the following two conditions:

$$0<\mu<\frac{2}{\lambda_{\max}}$$

$$\mu\sum_{k=0}^{p}\frac{\lambda_k}{2-\mu\lambda_k}<1$$

Note that the satisfaction condition of the above step size is required for the LMS algorithm to converge in the mean. The latter condition that μ satisfies guarantees that $\xi(\infty)$ is positive. Solving the steady-state value of Excess Mean Squared Error $\xi_{ex}(\infty)$ yields

$$\xi_{ex}(\infty)=\mu\xi_{\min}\frac{\mu\sum_{k=0}^{p}\frac{\lambda_k}{2-\mu\lambda_k}}{1-\mu\sum_{k=0}^{p}\frac{\lambda_k}{2-\mu\lambda_k}} \tag{6.29}$$

If $\mu\ll\frac{2}{\lambda_{\max}}$, then $\mu\lambda_k\ll 2$, there is

$$\frac{1}{2}\mu\sum_{k=0}^{p}\lambda_k<1$$

or

$$\mu<\frac{2}{tr(\boldsymbol{R}_x)}$$

When $\mu\ll\frac{2}{\lambda_{\max}}$, it follows that

$$\xi(\infty)\approx\xi_{\min}\frac{1}{1-\frac{1}{2}\mu tr(\boldsymbol{R}_x)}$$

and the Excess Mean Square Error in equation (6.29) is approximated by

$$\xi_{ex}(\infty)\approx\mu\xi_{\min}\frac{\frac{1}{2}tr(\boldsymbol{R}_x)}{1-\frac{1}{2}\mu tr(\boldsymbol{R}_x)}\approx\frac{1}{2}\mu\xi_{\min}tr(\boldsymbol{R}_x) \tag{6.30}$$

Therefore, for small step size μ, the excess mean square error is proportional to the step size, and the adaptive filters can be described in terms of their disorder, which is a normalized mean square error defined as follows

---

Definition: The misadjustment M is the ratio of the steady-state excess mean-square error to the minimum mean-square error

$$M=\frac{\xi_{ex}(\infty)}{\xi_{\min}}$$

From equation (6.30) we see that if the step size is small and satisfies $\mu\ll\frac{2}{\lambda_{\max}}$, then the misadjustment is approximately

$$M=\mu\frac{\frac{1}{2}tr(\boldsymbol{R}_x)}{1-\frac{1}{2}\mu tr(\boldsymbol{R}_x)}\approx\frac{1}{2}\mu tr(\boldsymbol{R}_x)$$

6.2.5 Normalized LMS

We can see that one of the difficulties in designing and implementing LMS adaptive filters is the choice of step size μ. For stationary processes, the LMS algorithm converges in the mean if $0<\mu<$

$2/\lambda_{\max}$, and converges in the mean-square if $0 < \mu < 2/tr(\boldsymbol{R}_x)$. However, since R_x is usually unknown, the $\lambda_{\max}$ or R_x must be estimated in order to use above boundary conditions. One way to solve this problem is to use the fact that, for stationary processes, $tr(\boldsymbol{R}_x) = (p+1)E\{|x(n)|^2\}$. Thus, the condition for mean-square convergence can be replaced with

$$0 < \mu < \frac{2}{(p+1)E\{|x(n)|^2\}}$$

where $E\{|x(n)|^2\}$ is the power of the process $x(n)$. This power can be estimated with a time average such as

$$\hat{E}\{|x(n)|^2\} = \frac{1}{p+1}\sum_{k=0}^{p}|x(n-k)|^2$$

which leads to the following constraint on the step size for the mean square convergence

$$0 < \mu < \frac{2}{\boldsymbol{x}^{\mathrm{H}}(n)\boldsymbol{x}(n)}$$

A quick way to apply this constraint to the LMS adaptive filter is to use a (time-varying) step size of the form

$$\mu(n) = \frac{\beta}{\boldsymbol{x}^{\mathrm{H}}(n)\boldsymbol{x}(n)} = \frac{\beta}{\|\boldsymbol{x}(n)\|^2} \tag{6.31}$$

where β is a normalized step size with $0 < \beta < 2$. The Normalized LMS algorithm (NLMS) is obtained by replacing μ in the LMS weight vector update equation with $\mu(n)$, which is given by

$$\boldsymbol{w}_{n+1} = \boldsymbol{w}_n + \beta\frac{\boldsymbol{x}^*(n)}{\|\boldsymbol{x}(n)\|^2}e(n) \tag{6.32}$$

In the LMS algorithm, the correction that is applied to $\boldsymbol{w}_n$ is proportional to the input vector $\boldsymbol{x}(n)$. Therefore, when $\boldsymbol{x}(n)$ is large, the LMS algorithm experiences a problem with gradient noise amplification. However, with the normalization of the LMS step size by $\|\boldsymbol{x}(n)\|^2$ in the NLMS algorithm, this noise amplification problem is diminished. Although the NLMS algorithm bypasses the problem of noise amplification, we are now faced with a similar problem that occurs when $\|\boldsymbol{x}(n)\|$ becomes too small. An alternative, therefore, is to use the following modification to the NLMS algorithm

$$\boldsymbol{w}_{n+1} = \boldsymbol{w}_n + \beta\frac{\boldsymbol{x}^*(n)}{\varepsilon + \|\boldsymbol{x}(n)\|^2}e(n) \tag{6.33}$$

where ε is a small positive number.

Compared with the LMS algorithms, it is easy to know that the normalized LMS algorithm (NLMS) is required to compute the normalization term $\|\boldsymbol{x}(n)\|^2$. The recursion of this term can be obtained as follows

$$\|\boldsymbol{x}(n+1)\|^2 = \|\boldsymbol{x}(n)\|^2 + |x(n+1)|^2 - |x(n-p)|^2$$

6.3 Affine Projection Adaptive Filter

6.3.1 Introduction to Affine Projection Algorithms

In mathematical terms, we may formulate the criterion for designing an affine projection filter as one of optimization subject to multiple constraints, as follows:

Minimize the squared Euclidean norm of the change in the weight vector

$$\delta\hat{\boldsymbol{w}}(n+1)=\hat{\boldsymbol{w}}(n+1)-\hat{\boldsymbol{w}}(n) \tag{6.34}$$

andsubject to the set of N constraints

$$d(n-k)=\hat{\boldsymbol{w}}^H(n+1)\boldsymbol{u}(n-k) \quad \text{for} \quad k=0,1,\cdots,N-1 \tag{6.35}$$

where N is smaller than the dimensionality M of the input data space or, equivalently, the weight space.

This constrained optimization criterion includes that of the normalized LMS algorithm as a special case, namely, $N=1$. We may view N, the number of constraints, as the order of the Affine Projection Adaptive filter.

Following the method of Lagrange multipliers with multiple constraints, we can combine Eq. (6.34) and (6.35) to set up the following cost function for the affine projection filter

$$J(n)= \|\hat{\boldsymbol{w}}(n+1)-\hat{\boldsymbol{w}}(n)\|^2+\sum_{k=0}^{N-1}\mathrm{Re}[\lambda_k^*(d(n-k)-\hat{\boldsymbol{w}}^H(n+1)\boldsymbol{u}(n-k))] \tag{6.36}$$

In this function, the λ_k are the Lagrange multipliers pertaining to the multiple constraints. For convenience of presentation, we introduce the following definitions:

An N-by-M data matrix $A(n)$ whose Hermitian transpose is defined by

$$A^H(n)=[\boldsymbol{u}(n),\boldsymbol{u}(n-1),\cdots,\boldsymbol{u}(n-N+1)] \tag{6.37}$$

An N-by-1 desired response vector whose Hermitian transpose is defined by

$$\boldsymbol{d}^H(n)=[d(n),d(n-1),\cdots,d(n-N+1)] \tag{6.38}$$

An N-by-1 Lagrange vector whose Hermitian transpose is defined by

$$\boldsymbol{\lambda}^H=[\lambda_0,\lambda_1,\cdots,\lambda_{N-1}] \tag{6.39}$$

Using these matrix definitions in Eq. (6.36), the cost function can be rewritten in the more compact form

$$J(n)= \|\hat{\boldsymbol{w}}(n+1)-\hat{\boldsymbol{w}}(n)\|^2+\mathrm{Re}[(\boldsymbol{d}(n)-A(n)\hat{\boldsymbol{w}}^H(n+1))^H\boldsymbol{\lambda}] \tag{6.40}$$

Then, according to the rules for the differentiation with respect to a complex-valued vector of the Wirtinger calculus, we find that the derivative of the cost function $J(n)$ with respect to the weight vector $\hat{\boldsymbol{w}}(n+1)$ is

$$\frac{\partial J(n)}{\partial\hat{\boldsymbol{w}}^H(n+1)}=2(\hat{\boldsymbol{w}}(n+1)-\hat{\boldsymbol{w}}(n))-\boldsymbol{A}^H(n)\boldsymbol{\lambda} \tag{6.41}$$

Setting this derivative equal to zero, we get

$$\delta\hat{\boldsymbol{w}}(n+1)=\frac{1}{2}\boldsymbol{A}^H(n)\boldsymbol{\lambda} \tag{6.42}$$

To eliminate the Lagrange vector $\boldsymbol{\lambda}$ from Eq. (6.42), we first use the definitions of Eqs. (6.37) and (6.38) to rewrite Eq. (6.35) in the equivalent form

$$\boldsymbol{d}(n)=\boldsymbol{A}(n)\hat{\boldsymbol{w}}(n+1) \tag{6.43}$$

Premultiplying both sides of Eq. (6.42) by $\boldsymbol{A}(n)$ and then using Eqs. (6.34) and (6.43) to eliminate the updated weight vector $\hat{\boldsymbol{w}}(n+1)$ yields

$$d(n)=A(n)\hat{w}(n)+\frac{1}{2}A(n)A^H(n)\lambda \tag{6.44}$$

from which we deduce the following:

• The difference between $d(n)$ and $A(n)\hat{w}(n)$, based on the data available at adaptation cycle n, is the N-by-1 error vector

$$e(n)=d(n)-A(n)\hat{w}(n) \tag{6.45}$$

• The matrix $A(n)A^H(n)$ is an N-by-N matrix with an inverse denoted by $(A(n)A^H(n))^{-1}$.

Thus, solving Eq. (6.44) for the Lagrange vector, we have

$$\lambda=2\ (A(n)A^H(n))^{-1}e(n) \tag{6.46}$$

Substituting this solution into Eq. (6.42) yields the optimum change in the weight vector, and we need to exercise control over the change in the weight vector from one adaptation cycle to the next, but keep the same direction. We do so by introducing the step-size parameter $\tilde{\mu}$, yielding that

$$\delta\hat{w}(n+1)=\tilde{\mu}\,A^H(n)(A(n)A^H(n))^{-1}e(n) \tag{6.47}$$

Finally we write the Affine Projection Adaptive filter algorithm

$$\hat{w}(n+1)=\hat{w}(n)+\tilde{\mu}A^H(n)(A(n)A^H(n))^{-1}e(n) \tag{6.48}$$

6.3.2 Affine Projection Operator

In this section, to determine this operator, we substitute Eq.(6.45) into Eq.(6.48), obtaining

$$\hat{w}(n+1)=[I-\tilde{\mu}A^H(n)(A(n)A^H(n))^{-1}A(n)]\hat{w}(n)+\tilde{\mu}A^H(n)(A(n)A^H(n))^{-1}d(n) \tag{6.49}$$

where I is the identity matrix. Define the projection operator

$$P=A^H(n)(A(n)A^H(n))^{-1}A(n) \tag{6.50}$$

which is uniquely determined by the data matrix$A(n)$. For prescribed $\tilde{\mu}$, $A(n)$, and $d(n)$, the complement projector $[I-\tilde{\mu}P]$ acts on the old weight vector $\hat{w}(n)$ to produce the updated weight vector $\hat{w}(n+1)$. Most importantly, it is the presence of the second term in Eq. (6.49), namely, $\tilde{\mu}A^H(n)(A(n)A^H(n))^{-1}d(n)$ that makes the complement projection into an affine projection rather than just a projection.

In Chapter 5, according to the method of Least Squares, we show that, for N less than M, the matrix $A^H(n)(A(n)A^H(n))^{-1}$ is the pseudoinverse of the data matrix $A(n)$. Using $A^+(n)$ to denote this pseudoinverse, we may simplify the updated formula of Eq. (6.49) as

$$\hat{w}(n+1)=[I-\tilde{\mu}A^+(n)A(n)]\hat{w}(n)+\tilde{\mu}A^+(n)d(n) \tag{6.51}$$

Indeed, it is because of the defining equation (6.51) that we may view the affine projection filter as an intermediate adaptive filter between the normalized LMS algorithm of Section 6.2.5 and the Recursive Least Squares (RLS) algorithm of Section 6.4, in terms of both computational complexity and performance.

6.3.3 Stability Analysis of the Affine Projection Adaptive Filter

As with the normalized LMS algorithm, we may base stability analysis of the Affine Projection Adaptive filter on the mean-square deviation $D(n)$. Subtracting Eq. (6.48) from the unknown weight vector $\boldsymbol{W}$ of the multiple regression model serving as a frame of reference, we may write

$$\varepsilon(n+1)=\varepsilon(n)-\tilde{\mu}\boldsymbol{A}^{H}(n)(\boldsymbol{A}(n)\boldsymbol{A}^{H}(n))^{-1}\boldsymbol{e}(n) \quad (6.52)$$

Rearranging and simplifying terms, we get theupdated equation of $D(n)$

$$\begin{aligned}D(n+1)-D(n)=&\tilde{\mu}^{2}E[\boldsymbol{e}^{H}(n)(^{A}(n)A^{H}(n))-1\boldsymbol{e}(n)]\\&-2\tilde{\mu}E\{Re[\zeta_{u}^{H}(n)(\boldsymbol{A}(n)\boldsymbol{A}^{H}(n))^{-1}\boldsymbol{e}(n)]\}\end{aligned} \quad (6.53)$$

where

$$\zeta_{u}(n)=\boldsymbol{A}(n)(\boldsymbol{w}(n)-\hat{\boldsymbol{w}}(n)) \quad (6.54)$$

is the undisturbed error vector. From Eq. (6.53) we readily see that the mean-square deviation $D(n)$ decreases monotonically with increasing n, provided that the step-size parameter $\tilde{\mu}$ satisfies the condition

$$0<\tilde{\mu}<\frac{2E\{\mathrm{Re}[\zeta_{u}^{H}(n)(\boldsymbol{A}(n)\boldsymbol{A}^{H}(n))^{-1}\boldsymbol{e}(n)]\}}{E[\boldsymbol{e}^{H}(n)(\boldsymbol{A}(n)\boldsymbol{A}^{H}(n))^{-1}\boldsymbol{e}(n)]} \quad (6.55)$$

which contains the corresponding formula of Eq. (6.33) for the normalized LMS algorithm as a special case. The optimal step size is defined by

$$\tilde{\mu}_{opt}=\frac{E\{\mathrm{Re}[\zeta_{u}^{H}(n)(\boldsymbol{A}(n)\boldsymbol{A}^{H}(n))^{-1}\boldsymbol{e}(n)]\}}{E[\boldsymbol{e}^{H}(n)(\boldsymbol{A}(n)\boldsymbol{A}^{H}(n))^{-1}\boldsymbol{e}(n)]} \quad (6.56)$$

6.4 RLS Adaptive Filter

6.4.1 Introduction to RLS

In each of the adaptive filtering methods discussed so far, we have considered gradient descent algorithms for the minimization of the mean-square error

$$\xi(n)=E\{|e(n)|^{2}\}$$

The difficulty with these methods is that they all require knowledge of the auto-correlation of the input process, $E\{x(n)x^{*}(n-k)\}$, and the cross-correlation between the input and the desired output, $E\{d(n)x^{*}(n-k)\}$. When this statistical information is unknown, we have been forced to estimate these statistics from the data. In the LMS adaptive filter, for example, these ensemble averages are estimated using instantaneous values,

$$\hat{E}\{e(n)x^{*}(n-k)\}=e(n)x^{*}(n-k) \quad (6.57)$$

Although this approach may be adequate in some applications, in others this gradient estimate may not provide a sufficiently rapid rate of convergence or a sufficiently small excess mean-square error. Therefore an alternative is to consider error measures that do not include expectations and that may be computed directly from the data. For example, a least squares error

$$\varepsilon(n)=\sum_{i=0}^{n}|e(i)|^{2}$$

requires no statistical information about $x(n)$ or $d(n)$, and may be evaluated directly from $x(n)$ and $d(n)$. However, there is an important philosophical difference between minimizing the least squares error and the mean-square error. Minimizing the mean-square error produces the same set of filter coefficients for all sequences that have the same statistics. Therefore, the coefficients do not depend on the incoming data, only on their ensemble average.

We are minimizing a squared error that depends explicitly on the specific values of $x(n)$ and $d(n)$ with the least squares error. Consequently, we get different filters for different signals. As a result, the filter coefficients that minimize the least squares error will be optimal for the given data rather than statistically optimal over a particular class of processes. In other words, different realizations of $x(n)$ and $d(n)$ will lead to different solutions, even if the statistics of these sequences are the same. In this section, we will look at the filters that are derived by minimizing a weighted least squares error, and derive an efficient algorithm for performing this minimization known as Recursive Least Squares.

6.4.2 Exponentially Weighted RLS

Let us reconsider the design of an FIR adaptive Wiener filter and find the filter coefficients

$$\boldsymbol{w}_n = [w_n(0), w_n(1), \cdots, w_n(p)]^{\mathrm{T}}$$

that minimize the weighted least squares error at time n.

$$\varepsilon(n) = \sum_{i=0}^{n} \lambda^{n-1} |e(i)|^2 \tag{6.58}$$

where $0 < \lambda \leqslant 1$ is an exponential weighting (forgetting) factor and

$$e(i) = d(i) - y(i) = d(i) - \boldsymbol{w}_n^{\mathrm{T}} \boldsymbol{x}(i) \tag{6.59}$$

Note that $e(i)$ is the difference between the desired signal $d(i)$ and the filtered output at time i, using the latest set of filter coefficients $w_n(k)$. Thus, in minimizing $\varepsilon(n)$, it is assumed that the weights $\boldsymbol{w}_n$ are held constant over the entire observation interval $[0, n]$. To find the coefficients that minimize $\varepsilon(n)$, we proceed exactly as we have done many times before by setting the derivative of $\varepsilon(n)$ with respect to $w_n^*(k)$ equal to zero for $k = 0, 1, \cdots, p$. Thus, we have

$$\frac{\partial \varepsilon(n)}{\partial w_n^*(k)} = -\sum_{i=0}^{n} \lambda^{n-i} e(i) x^*(i-k) = 0 \tag{6.60}$$

for $k = 0, 1, \cdots, p$. Incorporating equation (6.59) into equation (6.60) yields

$$\sum_{i=0}^{n} \lambda^{n-i} \{ d(i) - \sum_{l=0}^{p} w_n(l) x(i-l) \} x^*(i-k) = 0$$

Interchanging the order of summation and rearranging terms we have

$$\sum_{l=0}^{p} w_n(l) [\sum_{i=0}^{n} \lambda^{n-i} x(i-l) x^*(i-k)] = \sum_{i=0}^{n} \lambda^{n-i} d(i) x^*(i-k) \tag{6.61}$$

We may express these equations concisely in matrix form as follows

$$\boldsymbol{R}_x(n)\, \boldsymbol{w}_n = \boldsymbol{r}_{dx}(n) \tag{6.62}$$

where $\boldsymbol{R}_x(n)$ is a $(p+1) \times (p+1)$ exponentially weighted deterministic autocorrelation matrix for $x(n)$

$$\boldsymbol{R}_x(n) = \sum_{i=0}^{n} \lambda^{n-i}\, \boldsymbol{x}^*(i)\, \boldsymbol{x}^{\mathrm{T}}(i)$$

With $\boldsymbol{x}(i)$ the data vector

$$\boldsymbol{x}(i) = [x(i), x(i-1), \cdots, x(i-p)]^{\mathrm{T}}$$

and where $\boldsymbol{r}_{dx}(n)$ is the deterministic cross-correlation between $d(n)$ and $x(n)$,

$$\boldsymbol{r}_{dx}(n) = \sum_{i=0}^{n} \lambda^{n-i} d(i)\, \boldsymbol{x}^{*}(i) \tag{6.63}$$

Equation (6.62) is referred to as the deterministic normal equations.

Since $\boldsymbol{R}_x(n)$ and $\boldsymbol{r}_{dx}(n)$ both depend on n, instead of solving the deterministic normal equations directly, for each value of n, we will derive a recursive solution of the form

$$\boldsymbol{w}_n = \boldsymbol{w}_{n-1} + \Delta\, \boldsymbol{w}_{n-1}$$

where $\Delta \boldsymbol{w}_{n-1}$ is a correction that is applied to the solution at time $n-1$. Since

$$\boldsymbol{w}_n = \boldsymbol{R}_x^{-1}(n)\, \boldsymbol{r}_{dx}(n)$$

this recursion will be derived by first expressing $\boldsymbol{r}_{dx}(n)$ in terms of $\boldsymbol{r}_{dx}(n-1)$, and then deriving a recursion that allows us to evaluate $\boldsymbol{R}_x^{-1}(n)$ in terms of $\boldsymbol{R}_x^{-1}(n-1)$ and the new data vector $\boldsymbol{x}(n)$. From equation (6.63) it follows that the cross-correlation may be updated recursively as follows

$$\boldsymbol{r}_{dx}(n) = \lambda\, \boldsymbol{r}_{dx}(n-1) + d(n)\, \boldsymbol{x}^{*}(n) \tag{6.64}$$

Similarly, the autocorrelation matrix may be updated from $\boldsymbol{R}_x(n-1)$ and the new data vector $\boldsymbol{x}(n)$ using the recursion

$$\boldsymbol{R}_x(n) = \lambda\, \boldsymbol{R}_x(n-1) + \boldsymbol{x}^{*}(n)\, \boldsymbol{x}^{\mathrm{T}}(n) \tag{6.65}$$

However, since it is the inverse of $\boldsymbol{R}_x(n)$ that we are interested in, we may apply Woodbury's Identity, then we obtain the following recursion for the inversion of $\boldsymbol{R}_x(n)$

$$\boldsymbol{R}_x^{-1}(n) = \lambda^{-1} \boldsymbol{R}_x^{-1}(n-1) - \frac{\lambda^{-2} \boldsymbol{R}_x^{-1}(n-1)\, \boldsymbol{x}^{*}(n)\, \boldsymbol{x}^{\mathrm{T}}(n)\, \boldsymbol{R}_x^{-1}(n-1)}{1 + \lambda^{-1} \boldsymbol{x}^{\mathrm{T}}(n)\, \boldsymbol{R}_x^{-1}(n-1)\, \boldsymbol{x}^{*}(n)} \tag{6.66}$$

To simplify notation, we will let $\boldsymbol{P}(n)$ denote the inverse of the autocorrelation matrix at time n

$$\boldsymbol{P}(n) = \boldsymbol{R}_x^{-1}(n)$$

and define what is referred to as the gain vector $\boldsymbol{g}(n)$ as follows

$$\boldsymbol{g}(n) = \frac{\lambda^{-1} \boldsymbol{P}(n-1)\, \boldsymbol{x}^{*}(n)}{1 + \lambda^{-1} \boldsymbol{x}^{\mathrm{T}}(n) \boldsymbol{P}(n-1)\, \boldsymbol{x}^{*}(n)} \tag{6.67}$$

Incorporating these definitions into equation (6.66) we have

$$\boldsymbol{P}(n) = \lambda^{-1} [\boldsymbol{P}(n-1) - \boldsymbol{g}(n)\, \boldsymbol{x}^{\mathrm{T}}(n) \boldsymbol{P}(n-1)] \tag{6.68}$$

Finally, from equation (6.68) we see that the term multiplying $\boldsymbol{x}^{*}(n)$ is $\boldsymbol{g}(n)$ and we have

$$\boldsymbol{g}(n) = \boldsymbol{P}(n)\, \boldsymbol{x}^{*}(n) \tag{6.69}$$

To complete the recursion, we must derive the time-update equation for the coefficient vector $\boldsymbol{w}_n$. With

$$\boldsymbol{w}_n = \boldsymbol{P}(n)\, \boldsymbol{r}_{dx}(n)$$

it follows from the update for $\boldsymbol{r}_{dx}(n)$ given in equation (6.63) that

$$\boldsymbol{w}_n = \lambda \boldsymbol{P}(n)\, \boldsymbol{r}_{dx}(n-1) + d(n) \boldsymbol{P}(n)\, \boldsymbol{x}^{*} \tag{6.70}$$

Next, incorporating the update for $\boldsymbol{P}(n)$ given in equation (6.68) into the first term on the right side of equation (6.70), and setting $\boldsymbol{P}(n)\, \boldsymbol{x}^{*}(n) = \boldsymbol{g}(n)$ we have

$$\boldsymbol{w}_n = [\boldsymbol{P}(n-1) - \boldsymbol{g}(n)\, \boldsymbol{x}^{\mathrm{T}}(n) \boldsymbol{P}(n-1)]\, \boldsymbol{r}_{dx}(n-1) + d(n) \boldsymbol{g}(n)$$

Finally, recognizing that $\boldsymbol{P}(n-1)\boldsymbol{r}_{dx}(n-1)=\boldsymbol{w}_{n-1}$, it follows that

$$\boldsymbol{w}_n=\boldsymbol{w}_{n-1}+\boldsymbol{g}(n)[d(n)-\boldsymbol{w}_{n-1}^{\mathrm{T}}\boldsymbol{x}(n)]$$

which may be written as

$$\boldsymbol{w}_n=\boldsymbol{w}_{n-1}+\alpha(n)\boldsymbol{g}(n) \tag{6.71}$$

where

$$\alpha(n)=d(n)-\boldsymbol{w}_{n-1}^{\mathrm{T}}\boldsymbol{x}(n) \tag{6.72}$$

is the difference between $d(n)$ and the estimate of $d(n)$ that is formed by applying the previous set of filter coefficients $\boldsymbol{w}_{n-1}$ to the new data vector $\boldsymbol{x}(n)$. This sequence, called a priori error, is the error that would occur if the filter coefficients were not updated. The posteriori error, on the other hand, is the error that occurs after the weight vector is updated,

$$e(n)=d(n)-\boldsymbol{w}_n^{\mathrm{T}}\boldsymbol{x}(n)$$

Note that when $\alpha(n)$ is small, the current set of filter coefficients are close to their optimal values (in the Least Squares sense), and only a small correction needs to be applied to the coefficients. On the other hand, when $\alpha(n)$ is large, the current set of filter coefficients are not performing well in estimating $d(n)$ and a large correction must be applied to update the coefficients. The following Table 6.2 gives the summary of the Recursive Least Squares algorithm.

Table 6.2 The Summary of Recursive Least Squares

Parameters: p = Filter order λ = Exponential weighting factor δ = Value used to initialize $\boldsymbol{P}(0)$
Initialization: $\boldsymbol{w}_0=0$ $\boldsymbol{P}(0)=\delta^{-1}\boldsymbol{I}$
Computation: for $n=1,2,\cdots$ compute $z(n)=\boldsymbol{P}(n-1)\boldsymbol{x}^*(n)$ $\boldsymbol{g}(n)=\dfrac{1}{\lambda+\boldsymbol{x}^{\mathrm{T}}(n)z(n)}z(n)$ $\alpha(n)=d(n)-\boldsymbol{w}_{n-1}^{\mathrm{T}}\boldsymbol{x}(n)$ $\boldsymbol{w}_n=\boldsymbol{w}_{n-1}+\alpha(n)\boldsymbol{g}(n)$ $\boldsymbol{P}(n)=\dfrac{1}{\lambda}[\boldsymbol{P}(n-1)-\boldsymbol{g}(n)z^{\mathrm{H}}(n)]$

One final simplification may be realized if we note that, in the evaluation of the gain vector $\boldsymbol{g}(n)$, and the inverse autocorrelation matrix $\boldsymbol{P}(n)$, it is necessary to compute the product

$$z(n)=\boldsymbol{P}(n-1)\boldsymbol{x}^*(n) \tag{6.73}$$

Therefore, we may explicitly compute the filtered information vector and then use it in the calculation of both $\boldsymbol{g}(n)$ and $\boldsymbol{P}(n)$. We know that above Eqs (6.67), (6.68), (6.71), (6.72), and (6.73) form the exponentially weighted Recursive Least Squares (RLS) algorithm, which is summarized in Table 6.2. The special case of $\lambda=1$ is called as the growing window RLS algorithm.

The last issue that needs to be addressed is the initialization of the RLS algorithm. Since the recursion of the estimated vector $\boldsymbol{w}(n)$ and the inverse autocorrelation matrix $\boldsymbol{P}(n)$, the initialization for both of these terms are required. There are two ways to typically perform. The first is to build up the autocorrelation matrix recursively until it is full of rank ($p+1$ input vectors), and then compute the inverse matrix

$$P(0) = \left[\sum_{i=-p}^{0} \lambda^{-i} x^*(i)\, x^{\mathrm{T}}(i) \right]$$

In the same way of the inverse matrix $P(0)$, evaluate the cross-correlation vector

$$r_{dx}(0) = \sum_{i=-p}^{0} \lambda^{-i} d(i)\, x^*(i)$$

then we initialize w_0 by setting $w_0 = P(0) r_{dx}(0)$. The advantage of this approach is that optimality is preserved at each step since the RLS algorithm is initialized with the vector w_0 which minimizes the weighted least squares error $\varepsilon(0)$. However, it requires the direct inversion of $R_x(0)$, here we use the approach to initialize the autocorrelation matrix as follows

$$R_x(0) = \delta I$$

where δ is a small positive, i. e., $P(0) = \delta^{-1} I$, and the weight vector is initialized to be zero.

Unlike the LMS algorithm, which requires on the order of p multiplications and additions, the RLS algorithm requires on the order of p^2 operations. Specifically, the evaluation of $z(n)$ requires $(p+1)^2$ multiplications, computing the gain vector $g(n)$ requires $2(p+1)$ multiplications, finding the priori error $a(n)$ requires another $p+1$ multiplications, and the update of the inverse autocorrelation matrix $P(\mathrm{n})$ requires $2(p+1)^2$ multiplications for a total of $3(p+1)^2 + 3(p+1)$. There is a similar number of additions. However, what is gained with this increase in computational complexity over the LMS algorithm, is an increase in performance. Generally, the RLS algorithm converges faster than the LMS algorithm and is less sensitive to eigenvalue disparities in the autocorrelation matrix of $x(n)$ for stationary processes. On the other hand, without exponential weighting, RLS does not perform very well in tracking nonstationary processes. This is due to the fact that, with $\lambda = 1$, all of the data is equally weighted in estimating the correlations. Although exponential weighting improves the tracking characteristics of RLS, it is not clear how to choose λ, and, in some cases, the LMS algorithm may have better tracking properties.

6. 5 Summary

In this chapter, we looked at a number of techniques for processing nonstationary signals using adaptive filters. These techniques have been extensively used in a variety of applications including system identification, signal modeling, spectrum estimation, noise cancellation, and adaptive equalization. Our discussion of adaptive filtering began with the design of an FIR filter to minimize the mean-square error $\xi(n) = E\{\,|e_{(n)}|^2\}$ between a desired process $d(n)$ and an estimate of this process that is formed by filtering another process $x(n)$. First, we considered a steepest descent approach to minimize $\xi(n)$. However, since the gradient of $\xi(n)$ involves an expectation of $e(n) * x(n)$, this approach requires knowledge of the statistics of $x(n)$ and $d(n)$, therefore, it has limited use in practice. Next, we replaced the ensemble average $E\{\,|e_{(n)}|^2\}$ with the instantaneous squared error $|e_{(n)}|^2$. This led to the LMS algorithm, a simple and often effective algorithm that does not require any ensemble averages to be known. For wide-sense stationary processes, the LMS algorithm converges in the mean if the step size is positive and no larger than $2/\lambda_{\max}$, and it converges in the meansquare if the step size is positive and no larger than $2/tr(\boldsymbol{R}_x)$. We then looked at the normalized LMS algorithm which simplifies the selection of the step size to ensure that the coefficients converge.

The affine projection filter is a generalization of the normalized LMS algorithm. Specifically, the adjustment term $\beta \dfrac{x^*(n)}{\varepsilon + \| x(n) \|^2} e(n)$ applied to the tap-weight vector of the normalized LMS algorithm is replaced by the more elaborate term $\tilde{\mu} A^H(n)(A(n)A^H(n)+\delta I)^{-1} e(n)$, where I is the identity matrix, and δ is a small positive constant. Because of the reuse of $(N-1)$ past values of both the input vector $u(n)$ and the desired response $d(n)$, the affine projection filter may be viewed as an adaptive filter that is intermediate between the normalized LMS algorithms and the RLS algorithms. As a result, the affine projection filter provides a significant improvement in convergence, which unfortunately comes at the cost of increased computational complexity.

This chapter also concluded a derivation of the Recursive Least Squares algorithms to minimize the deterministic least squares error. Although its computation is more complex than the LMS algorithm, its convergence is quiet rapid. In order to effectively track a nonstationary process, it is necessary to use the exponentially weighted RLS algorithm.

Exercises

1. First, we consider examples of algorithms for adaptive filtering: the affine projection algorithm (APA), the LMS algorithm, and the RLS algorithm. It is required to be able to derive algorithm iteration formulas based on what we have learned and to implement Matlab program simulations with some simulation parameter setting as follows.

Input signal: generated by Gaussian white noise with mean 0 and variance 1 through the following AR model:

$$H(z) = 0.1 - 0.2z^{-1} - 0.3z^{-2} + 0.4z^{-3} + 0.4z^{-4} - 0.2z^{-5} - 0.1z^{-6}$$

The order of the Affine Projection Adaptive filter: T=4

Transmission channel: $W_1(z) = \sum_{n=0}^{19} z^{-n} - \sum_{n=21}^{M-1} z^{-n}$ or $W_2(z) = -\sum_{n=0}^{M-1} z^{-n}$, where M=40

Signal-to-noise ratio: 30 dB

Step factor: 0.20

Data length: 5000

Number of runs: 100 independent runs

Based on the above simulation settings, complete the following problems.

(1): Compare the convergence learning curve graphs of the affine projection algorithm, the LMS algorithm and the RLS algorithm, and draw meaningful conclusions.

(2): Change the order of the affine projection to 4, 8, 12 and compare the convergence and steady-state performance of the APA algorithm at different orders.

(3): Change the signal-to-noise ratio to 15dB, 30dB and 45dB respectively, and compare the convergence and steady-state performance of APA algorithm under different signal-to-noise ratios.

Experimental results: mean square deviation (MSD) is introduced as a measure of convergence and steady-state performance, and the MSD learning curve is plotted.

$$MSD = \lim_{i\to\infty} E\,\|\hat{w}_i - w_0\|^2$$

Solution:

some relevant Matlab source code:

LMS:

```
w0=[1;1;1;1;1;1;1;1;1;1;1;1;1;1;1;1;1;1;1;1;-1;-1;-1;-1;-1;-1;-1;-1;-1;-1;
-1;-1;-1;-1;-1;-1;-1;-1;-1;-1];
    wlm0=zeros(n,1);
    wlm1=zeros(n,1);
    e=zeros(1,itr);
    w_MSE2=zeros(1,itr-39);
    for k=n:itr;
        xl=xn(1,k:-1:k-n+1)';
         xl=x(1,k:-1:k-n+1)';
        y=wlm0' * xl;
        d=xl' * w0+noise(k);
        e=d-y;
        m=0.01;
        wlm1=wlm0+2 * m * e * xl;
        wlm0=wlm1;
        w_MSE2(k-39)=norm(wlm0-w0);
    end
    Hw_MSE2=Hw_MSE2+w_MSE2;
```

APA:

```
w0=[1;1;1;1;1;1;1;1;1;1;1;1;1;1;1;1;1;1;1;1;-1;-1;-1;-1;-1;-1;-1;-1;-1;-1;
-1;-1;-1;-1;-1;-1;-1;-1;-1;-1];
Hwlm1=zeros(n,itr-n-p+2);
    Xnp=zeros(p,n);
    Y=zeros(itr,1);
    w_MSE1=zeros(1,itr-n-p+2);
    wlm0=zeros(n,1);
    Y=zeros(itr,1);
    D=zeros(p,1);
    E=zeros(p,1);
```

```
for k=p+n-1:itr
        for i=1:p
            for j=1:n
            Xnp(i,j)=xn(k+2-i-j);
                Xnp(i,j)=x(k+2-i-j);
            end
        end
        D=Xnp * w0+noise(k:-1:k-p+1)';
        Y(k:-1:k-p+1)=Xnp * wlm0;
        E=D-Y(k:-1:k-p+1);
        sita=0.001;
        I=eye(p);
        wlm0=wlm0+mu * Xnp' * pinv(sita * I+Xnp * Xnp') * E;
        w_MSE1(k-n-p+2)=norm(wlm0-w0);
        Hwlm1(:,k-n-p+2)=wlm0;
end
```

RLS:

```
noise=sqrt(0.01) * randn(1,itr);
w0=[1;1;1;1;1;1;1;1;1;1;1;1;1;1;1;1;1;1;1;1;-1;-1;-1;-1;-1;-1;-1;-1;-1;-1;
-1;-1;-1;-1;-1;-1;-1;-1;-1;-1];
wrl0=zeros(M,1);
wrl1=zeros(M,1);
e=zeros(itr,1);
w_MSE=zeros(1,itr-39);
Hwrl=zeros(M,itr-39);
P=10^6 * eye(M,M);
for n=M:itr
xr=xn(n:-1:n-M+1)';
pi_=xr' * P;
k=Lam+pi_ * xr;
K=pi_'/k;
d=w0' * xr+noise(n-39);
e=d-wrl0' * xr;
wrl1=wrl0+K * e;
PPrime=K * pi_;
P=P-PPrime;
wrl0=wrl1;
w_MSE(n-39)=norm(wrl1-w0);
end
```

2. In communications over phone lines, a signal travelling from a far-end point to a near-end point is usually reflected at the near-end due to mismatches in circuitry (e. g, , hybrid connections). The reflected signal travels back to the far-end point in the form of an echo. As a result, the speaker at the far-end receives, in addition to the desired signal from the near-end speaker, an attenuated replica of his own signal in the form of an echo-see Figure 6. 2.

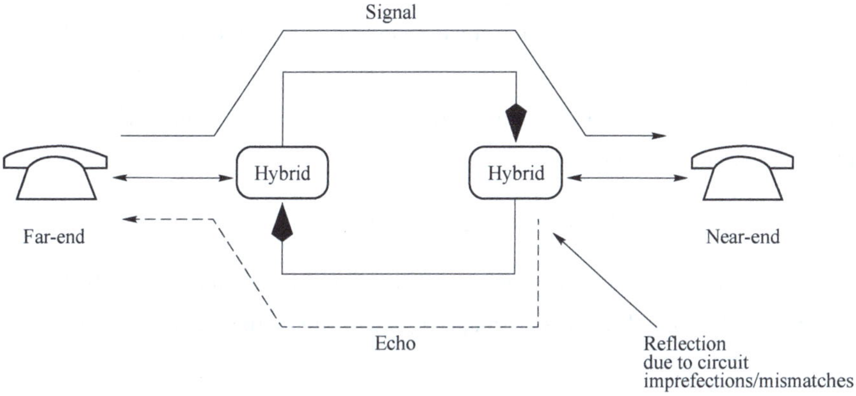

Figure 6. 2 The signal at the far-end is reflected at the near-end due to circuit mismatches and travels back to the far-end

The echo interferes with the quality of the received signal. A common way to provide better voice quality at both ends is to employ adaptive line echo cancellers (LEC). At the near-end, for example, the signal feeding the LEC is the far-end signal while the reference signal is its reflected version-see Figure 6. 3. In the figure, the output of the adaptive LEC generates a replica of the echo, and the error signal is therefore a "clean" signal that is transmitted to the far-end. The signals in this project are assumed to be sampled at 8 kHz.

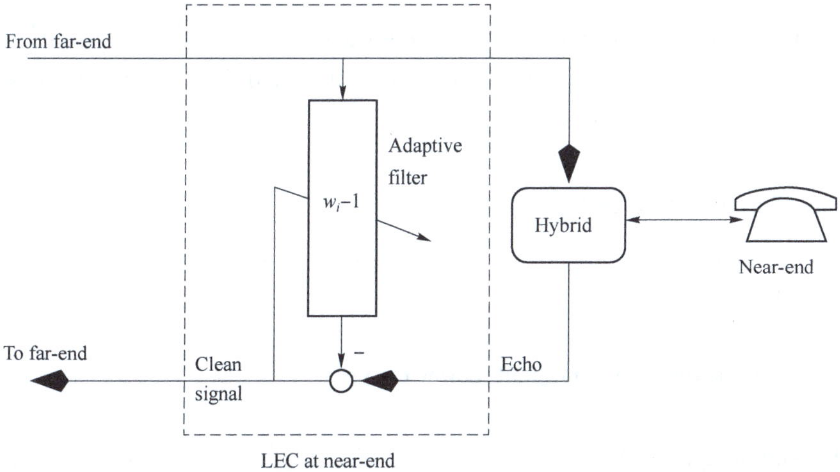

Figure 6. 3 An adaptive line echo canceller at the near-end

(1): Use the Matlab to generate an echo path, and save as the 'path. mat', load the file path. mat, which contains the impulse response sequence of a typical echo path. Plot the impulse and frequency responses of the echo path.

(2): Use the Matlab to generate a speech source input signal, and save as the 'css. mat', load the file css. mat, which contains 5,600 samples of a composite source signal; it is a synthetic signal that emulates the properties of speech. Specifically, it contains segments of pause, segments of periodic excitation and segments with white-noise properties. Plot the samples of the CSS data, as well as their spectrum.

(3): Concatenate five such blocks and feed them into the echo path. Plot the resulting echo signal. Estimate the input and output powers in dB using

$$\hat{P} = 10\log_{10}\left(\frac{1}{N}\sum_{i=1}^{N} |\text{signal}(i)|^2\right)$$

where N denotes the length of the sequence. Evaluate the attenuation in dB that is introduced by the echo path as the signal travels through it; this attenuation is called the echo-retum-loss (ERL).

(4): Use 10 blocks of CSS data as far-end signal, and the corresponding output of the echo path as the echo signal. Choose an adaptive line echo canceller with 128 taps. Train the canceller by using as input data the far-end signal, i. e., $u(i) = \text{far\_end}(i)$, and as reference data the echo signal, i. e., $d(i) = \text{echo}(i)$. Use NLMS with $\varepsilon = 10^{-6}$ and $\mu = 0.25$. Plot the far-end signal, the echo, and the error signal provided by the adaptive filter. Plot also the echo path and its estimate by the adaptive filter at the end of the simulation.

(5): Estimate the steady-state power of the error signal and measure its attenuation in dB relative to the echo signal. Use the last 5600 of the signals to estimate their powers. The difference in power is a measure of the attenuation introduced by the LEC and it is called the echo-return-loss-enhancement (ERLE).

(6): Fix the input power at 0 dB and add white Gaussian noise with variance $\sigma_v^2 = 0.000,1$ to the echo signal. Train the LEC using 80 blocks of CSS data and measure the steady-state ERLE. Compare the simulated and theoretical ERLEs.

Solution:

some relevant Matlab source code:

```
far_end_unit_power=[ ];
for i=1:10
        far_end_unit_power=[far_end_unit_power css];
end
N=length(far_end_unit_power);
M=128;
mu=0.25;
epsilon=1e-6;
```

```
far_end=far_end_unit_power;
echo=filter(path,1,far_end);
w=zeros(M,1); % weight estimate
u=zeros(1,M); % regressor
disp('Please be patient...This takes a while...')
for i=1:N
    u=[far_end(i) u(1:M-1)];
    e(i)=echo(i)-u*w;
    factor=epsilon+(norm(u)^2);
    w=w+(mu/ factor)*u'*e(i);
end
Pecho=10*log10((norm(echo(N-5600:N))^2/ 5600));
Perror=10*log10((norm(e(N-5600:N))^2)/ 5600);
ERLE=Pecho-Perror
```

References

[1] S. T. Alexander, *Adaptive Signal Processing: Theory and Applications*, Springer-Verlag, New York, 1986.

[2] Haykin S. *Adaptive Filter Theory*. 4th ed. Upper Saddle River: Prentice Hall, 2002.

[3] Dinjz, P. S. R. Adaptive Filtering: Algorithms and Practical Implementation, Kluwer Academic Publishers: Boston, 1997.

[4] N. J. Bershad, "Analysis of the normalized LMS algorithm with Gaussian inputs," *IEEE Trans. Acoust.*, *Speech*, *Sig. Proc.*, vol. ASSP-34, pp. 793-806, 1986.

[5] L. Li, J. A. Chambers, C. G. Lopes and A. H. Sayed, "Distributed Estimation Over an Adaptive Incremental Network Based on the Affine Projection Algorithm," in *IEEE Transactions on Signal Processing*, vol. 58, no. 1, pp. 151-164, Jan. 2010.